Henry Lee

Sea Monsters Unmasked

Henry Lee

Sea Monsters Unmasked

ISBN/EAN: 9783337397043

Printed in Europe, USA, Canada, Australia, Japan

Cover: Foto ©berggeist007 / pixelio.de

More available books at **www.hansebooks.com**

International Fisheries Exhibition
LONDON, 1883

SEA MONSTERS UNMASKED

BY

HENRY LEE, F.L.S., F.G.S., F.Z.S.
SOMETIME NATURALIST OF THE BRIGHTON AQUARIUM

AND

AUTHOR OF 'THE OCTOPUS, OR THE DEVIL-FISH OF FICTION AND FACT'

ILLUSTRATED

LONDON
WILLIAM CLOWES AND SONS, Limited
INTERNATIONAL FISHERIES EXHIBITION
AND 13 CHARING CROSS, S.W.
1883

PREFACE.

As I commence this little history of two sea monsters there comes to my mind a remark made to me by my friend, Mr. Samuel L. Clemens—"Mark Twain"—which illustrates a feeling that many a writer must have experienced when dealing with a subject that has been previously well handled. Expressing to me one day the gratification he felt in having made many pleasant acquaintances in England, he added, with dry humour, and a grave countenance, "Yes! I owe your countrymen no grudge or ill-will. I freely forgive them, though one of them did me a grievous wrong, an irreparable injury! It was Shakspeare: if he had not written those plays of his, I should have done so! They contain *my* thoughts, *my* sentiments! He forestalled me!"

In treating of the so-called "sea-serpent," I have been anticipated by many able writers. Mr. Gosse, in his delightful book, 'The Romance of Natural History,' published in 1862, devoted a chapter to it; and numerous articles concerning it have appeared in various papers and periodicals.

But, for the information from which those authors have drawn their inferences, and on which they have founded their opinions, they have been greatly indebted, as must be all who have seriously to consider this subject, to the late experienced editor of the *Zoologist*, Mr. Edward Newman, a man of wonderful power of mind, of great judgment, a profound thinker, and an able writer. At a time when, as he said, "the shafts of ridicule were launched against believers and unbelievers in the sea-serpent in a very pleasing and impartial manner," he, in the true spirit of philosophical inquiry, in 1847, opened the columns of his magazine to correspondence on this topic, and all the more recent reports of marine monsters having been seen are therein recorded. To him, therefore, the fullest acknowledgments are due.

The great cuttles, also, have been the subject of articles in various magazines, notably one by Mr. W. Saville Kent, F.L.S., in the 'Popular Science Review' of April, 1874, and a chapter in my little book on the Octopus, published in 1873, is also devoted to them. In writing of them as the living representatives of the kraken, and as having been frequently mistaken for the "sea-serpent," my deductions have been drawn from personal knowledge, and an intimate acquaintance with the habits, form, and structure of the animals described. It was only by watching the movements of specimens of the "common squid" (*Loligo vulgaris*), and the "little squid" (*L. media*), which lived in the tanks of the Brighton Aquarium, that I recognised in their peculiar habit of occasionally swimming half-submerged, with uplifted caudal extremity, and trailing arms, the fact that I had before me the "sea-serpent" of many a well-authenticated anecdote. A mere knowledge of their form and anatomy after death had never suggested to me that which became at once apparent when I saw them in life.

It is a pleasure to me to acknowledge gratefully the kindness I have met with in connection with the illustrations of this book. The proprietors of the *Illustrated London News* not only gave me permission to copy, in reduced size, their two pictures of the *Dædalus* incident, but presented to me electrotype copies of all others small enough for these pages—namely, "Jonah and the Monster," Egede's "Sea-Serpent," and the Whale as seen from the *Pauline*. Equally kind have been the proprietors of the *Field*. To them I am greatly indebted for their permission to copy the beautiful woodcuts of the "Octopus at Rest," "The Sepia seizing its Prey," and the arms of the Newfoundland squids, and also for "electros" of the two curious Japanese engravings, all of which originally appeared in their paper. From the *Graphic* I have had similar permission to copy any cuts that might be thought suitable, and the illustrations of the sea-serpent, as seen from Her Majesty's yacht *Osborne* and the *City of Baltimore*, are from that journal. Messrs. Nisbet most courteously allowed me to have a copy of the block of the *Enaliosaurus* swimming, which was one of the numerous pictures in Mr. Gosse's book, published by them, already referred to. And last, not least, I have to thank Miss Ellen Woodward, daughter of my friend, Dr. Henry Woodward, F.R.S., for enabling me to better explain the movements and appearances of the squids when swimming, and when raising their bodies out of water in an erect position, by carefully drawing them from my rough sketches.

HENRY LEE.

SAVAGE CLUB;
July 21st, 1883.

LIST OF ILLUSTRATIONS.

Frontispiece.—The Sea Serpent as first seen from H.M.S. *Dædalus.*

FIG.		PAGE
1.	Beak and Arms of a Decapod Cuttle	16
2.	The Octopus (*Octopus vulgaris*)	18
3.	The Cuttle (*Sepia officinalis*)	21
4.	Hooked Tentacles of *Onychoteuthis*	23
5.	Japanese fisherman attacked by a Cuttle	29
6.	Arms of a great Cuttle exhibited in a Japanese fish-shop	29
7.	Facsimile of De Montfort's "*Poulpe colossal*"	32
8.	Gigantic Calamary caught by the French despatch vessel *Alecton*, near Teneriffe	39
9.	Tentacle of a great Calamary (*Architeuthis princeps*) taken in Conception Bay, Newfoundland	43
10.	Head and Tentacles of a great Calamary (*Architeuthis princeps*) taken in Logie Bay, Newfoundland	44
11.	Jonah and the Sea Monster	55
12.	Sea Serpent seizing a man on board ship	58
13.	Gigantic Lobster dragging a man from a ship	58
14.	Pontoppidan's "Sea Serpent"	63
15.	The Animal drawn by Mr. Bing as having been seen by Egede	66
16.	The Animal which Egede probably saw	67
17.	The Sea Serpent of the Wernerian Society (*facsimile*)	69
18.	A Calamary swimming at the surface of the sea	77
19.	The Sea Serpent passing under the quarter of H.M.S. *Dædalus*	81
20.	The Sea Serpent and Sperm Whale as seen from the *Pauline*	91
21.	The Sea Serpent as seen from the *City of Baltimore*	93
22.	The Sea Serpent as seen from H.M. yacht *Osborne*. Phase 1	94
23.	The Sea Serpent as seen from H.M. yacht *Osborne*. Phase 2	94
24.	Skeleton of the *Plesiosaurus*, restored by Mr. Conybeare	98
25.	The Sea Serpent on the Enaliosaurian hypothesis	100

SEA MONSTERS UNMASKED.

THE KRAKEN.

IN the legends and traditions of northern nations, stories of the existence of a marine animal of such enormous size that it more resembled an island than an organised being frequently found a place. It is thus described in an ancient manuscript (about A.D. 1180), attributed to the Norwegian King Sverre; and the belief in it has been alluded to by other Scandinavian writers from an early period to the present day. It was an obscure and mysterious sea-monster, known as the Kraken, whose form and nature were imperfectly understood, and it was peculiarly the object of popular wonder and superstitious dread.

Eric Pontoppidan, the younger, Bishop of Bergen, and member of the Royal Academy of Sciences at Copenhagen, is generally, but unjustly, regarded as the inventor of the semi-fabulous Kraken, and is constantly misquoted by authors who have never read his work,* and who, one after another, have copied from their predecessors erroneous statements concerning him. More than half a century before him, Christian Francis Paullinus,† a physician and naturalist of Eisenach, who evinced in his writings an admiration of

---

* 'Natural History of Norway.' A.D. 1751.
† Born 1643; died 1712.

the marvellous rather than of the useful, had described as resembling Gesner's 'Heracleoticon,' a monstrous animal which occasionally rose from the sea on the coasts of Lapland and Finmark, and which was of such enormous dimensions, that a regiment of soldiers could conveniently manœuvre on its back. About the same date, but a little earlier, Bartholinus, a learned Dane, told how, on a certain occasion, the Bishop of Midaros found the Kraken quietly reposing on the shore, and mistaking the enormous creature for a huge rock, erected an altar upon it and performed mass. The Kraken respectfully waited till the ceremony was concluded, and the reverend prelate safe on shore, and then sank beneath the waves.

And a hundred and fifty years before Bartholinus and Paullinus wrote, Olaus Magnus,[*] Archbishop of Upsala, in Sweden, had related many wondrous narratives of sea-monsters,—tales which had gathered and accumulated marvels as they had been passed on from generation to generation in oral history, and which he took care to bequeath to his successors undeprived of any of their fascination. According to him, the Kraken was not so polite to the laity as to the Bishop, for when some fishermen lighted a fire on its back, it sank beneath their feet, and overwhelmed them in the waters.

Pontoppidan was not a fabricator of falsehoods; but, in

---

[*] Olaus Magnus has sometimes been mistaken for his brother and predecessor in the archiepiscopal see, Johan Magnus, author of a book entitled 'Gothorum, Suevorumque Historia.' Olaus was the last Roman Catholic archbishop of the Swedish church, and when the Reformation, supported by Gustavus Vasa, gained the ascendancy in Sweden, he remained true to his faith, and retired to Rome, where he wrote his work, 'Historia de Gentibus Septentrionalibus,' Romæ, 1555. An English translation of this book was published by J. Streater, in 1658. It does not contain the illustrations.

collecting evidence relating to the "great beasts" living in "the great and wide sea," was influenced, as he tells us, by "a desire to extend the popular knowledge of the glorious works of a beneficent Creator." He gave too much credence to contemporary narratives and old traditions of floating islands and sea monsters, and to the superstitious beliefs and exaggerated statements of ignorant fishermen: but if those who ridicule him had lived in his day and amongst his people, they would probably have done the same; for even Linnæus was led to believe in the Kraken, and catalogued it in the first edition of his 'Systema Naturæ,' as '*Sepia Microcosmos.*' He seems to have afterwards had cause to discredit his information respecting it, for he omitted it in the next edition. The Norwegian bishop was a conscientious and painstaking investigator, and the tone of his writings is neither that of an intentional deceiver nor of an incautious dupe. He diligently endeavoured to separate the truth from the cloud of error and fiction by which it was obscured; and in this he was to a great extent successful, for he correctly identifies, from the vague and perplexing descriptions submitted to him, the animal whose habits and structure had given rise to so many terror-laden narratives and extravagant traditions.

The following are some of his remarks on the subject of this gigantic and ill-defined animal. Although I have greatly abbreviated them, I have thought it right to quote them at considerable length, that the modest and candid spirit in which they were written may be understood:[*]

"Amongst the many things," he says, "which are in the ocean, and concealed from our eyes, or only presented to our view for a few minutes, is the Kraken. This creature is the largest and most surprising of all the animal creation, and consequently well de-

[*] 'Natural History of Norway,' vol. ii., p. 210.

serves such an account as the nature of the thing, according to the Creator's wise ordinances, will admit of. Such I shall give at present, and perhaps much greater light on this subject may be reserved for posterity.

"Our fishermen unanimously affirm, and without the least variation in their accounts, that when they row out several miles to sea, particularly in the hot summer days, and by their situation (which they know by taking a view of different points of land) expect to find eighty or a hundred fathoms of water, it often happens that they do not find above twenty or thirty, and sometimes less. At these places they generally find the greatest plenty of fish, especially cod and ling. Their lines, they say, are no sooner out than they may draw them up with the hooks all full of fish. By this they know that the Kraken is at the bottom. They say this creature causes those unnatural shallows mentioned above, and prevents their sounding. These the fishermen are always glad to find, looking upon them as a means of their taking abundance of fish. There are sometimes twenty boats or more got together and throwing out their lines at a moderate distance from each other; and the only thing they then have to observe is whether the depth continues the same, which they know by their lines, or whether it grows shallower, by their seeming to have less water. If this last be the case they know that the Kraken is raising himself nearer the surface, and then it is not time for them to stay any longer; they immediately leave off fishing, take to their oars, and get away as fast as they can. When they have reached the usual depth of the place, and find themselves out of danger, they lie upon their oars, and in a few minutes after they see this enormous monster come up to the surface of the water; he there shows himself sufficiently, though his whole body does not appear, which, in all likelihood, no human eye ever beheld. Its back or upper part, which seems to be in appearance about an English mile and a half in circumference (some say more, but I chuse the least for greater certainty), looks at first like a number of small islands surrounded with something that floats and fluctuates like sea-weeds. Here and there a larger rising is observed like sand-banks, on which various kinds of small fishes are seen continually leaping about till they roll off into the water from the sides of it; at last several bright points or horns appear, which grow thicker and thicker the higher they rise above the surface of the water, and sometimes they stand up as high and as large as the masts of middle-sized vessels. It seems these are the creature's arms, and it is said if they were to lay hold of the largest man of war they would pull it down to the bottom. After this monster has been on the surface of the water a short time it begins slowly to sink again, and then the danger is as great as before; because the motion of his sinking causes such a swell in the sea, and such an eddy or whirlpool, that it draws everything down with it, like the current of the river Male.

"As this enormous sea-animal in all probability may be reckoned of the Polype, or of the Starfish kind, as shall hereafter be more fully proved, it seems that the parts which are seen rising at its pleasure, and are called arms, are properly the tentacula, or feeling instruments, called horns, as well as arms. With these they move themselves, and likewise gather in their food.

"Besides these, for this last purpose the great Creator has also given this creature a strong and peculiar scent, which it can emit at certain times, and by means of which it beguiles and draws other fish to come in heaps about it. This animal has another strange property, known by the experience of many old fishermen. They observe that for some months the Kraken or Krabben is continually eating, and in other months he always voids his excrements. During this evacuation the surface of the water is coloured with the excrement, and appears quite thick and turbid. This muddiness is said to be so very agreeable to the smell or taste of other fishes, or to both, that they gather together from all parts to it, and keep for that purpose directly over the Kraken; he then opens his arms or horns, seizes and swallows his welcome guests, and converts them after due time, by digestion, into a bait for other fish of the same kind. I relate what is affirmed by many; but I cannot give so certain assurances of this particular, as I can of the existence of this surprising creature; though I do not find anything in it absolutely contrary to Nature. As we can hardly expect to examine this enormous sea-animal alive, I am the more concerned that nobody embraced that opportunity which, according to the following account once did, and perhaps never more may offer, of seeing it entire when dead."

The lost opportunity which the worthy prelate thus lamented, with the true feeling of a naturalist, was made known to him by the Rev. Mr. Friis, Consistorial Assessor, Minister of Bodoen in Nordland, and Vicar of the college for promoting Christian knowledge, and was to the following effect :

"In the year 1680, a Krake (perhaps a young and foolish one) came into the water that runs between the rocks and cliffs in the parish of Alstaboug, though the general custom of that creature is to keep always several leagues from land, and therefore of course they must die there. It happened that its extended long arms or antennæ, which this creature seems to use like the snail in turning about, caught hold of some trees standing near the water, which might easily have been torn up by the roots; but beside this, as it was found afterwards, he entangled himself in some openings or clefts in the rock, and therein stuck so fast, and hung so unfortunately, that he could not work himself out, but perished and putrefied on the spot. The carcass, which was a long while decaying, and filled great part of that narrow channel, made it almost impassable by its intolerable stench.

"The Kraken has never been known to do any great harm, except," the Author quaintly says, " they have taken away the lives of those who consequently could not bring the tidings. I have heard but one instance mentioned, which happened a few years ago, near Fridrichstad, in the diocess of Aggerhuus. They say that two fishermen accidentally, and to their great surprise, fell into such a spot on the water as has been before described, full of a thick slime almost like a morass. They immediately strove to get out of this place, but they had not time to turn quick enough to save themselves from one of the Kraken's horns, which crushed the head of the boat, so that it was with great difficulty they saved their lives on the wreck, though the weather was as calm as possible; for these monsters, like the sea-snake, never appear at other times."

Pontoppidan then reviews the stories of floating islands which suddenly appear, and as suddenly vanish, commonly credited, and especially mentioned by Luke Debes in his 'Description of Faroe.'

"These islands in the boisterous ocean could not be imagined," he says, "to be of the nature of real floating islands, because they could not possibly stand against the violence of the waves in the ocean, which break the largest vessels, and therefore our sailors have concluded this delusion could come from no other than the great deceiver, the devil."

This accusation, the good bishop, in his desire to be strictly impartial, will not admit on such hear-say evidence, but is determined to, literally, "give the devil his due;" for he warns his readers that "we ought not to charge that apostate spirit without a cause; for," he adds, "I rather think that this devil who so suddenly makes and unmakes these floating islands, is nothing else but the Kraken."

Referring to a monster described by Pliny, he repeats his belief that "This sea-animal belongs to the Polype, or Star-fish species;" but he becomes very much "mixed" between the *Cephalopoda* and the *Asteridæ*, between the pedal segments, or arms, of the cuttle radiating from its head, and the rays of a Star-fish radiating from a central portion of the body. He evidently inclines strongly towards a particular Star-fish, the rays of which continually divide and subdivide themselves, or, as he describes it, "which shoots its rays into branches like those of trees," and to which he gave the name of "Medusa's Head," a title by which, in its Greek form, *Gorgonocephalus*, it is still known to zoologists. "These Medusa's Heads," he says, "are supposed by some seafaring people here, to be the young of the Sea-Krake; perhaps they are its smallest ovula." After considering other reports concerning the Kraken, he arrives at the following definite opinion:

"We learn from all this that the Polype or Starfish have amongst their various species some that are much larger than others; and, according to all appearance, amongst the very largest inhabitants of the ocean. If the axiom be true that greatness or littleness makes no change in the species, then this Krake must be of the Polypus kind, notwithstanding its enormous size."

His diagnosis is correct; but it is stated with a modesty which his detractors would do well to imitate; and his concluding words on this subject place him in a light very different from that in which he is popularly regarded:

"I do not in the least insist on this conjecture being true," he writes, "but willingly submit my suppositions in this and every other dubious matter to the judgment of those who are better experienced. If I was an admirer of uncertain reports and fabulous stories, I might here add much more concerning this and other Norwegian sea-monsters, whose existence I will not take upon me to deny, but do not chuse, by a mixture of uncertain relations to make such account appear doubtful as I myself believe to be true and well attested. I shall therefore quit the subject here, and leave it to future writers on this plan to complete what I have imperfectly sketched out, by further experience, which is always the best instructor.'"

It is easy to recognise in Pontoppidan's description of the Kraken, the form and habits of one of the "Cuttle-fishes," so-called. The appearance of its numerous arms, with which it gathers in its food, and which grow thicker and thicker as they rise above the surface, is just what would take place in the case of one of the pelagic species of these mollusks raising its head out of the sea. The rendering of the water turbid and thick by the emission of a substance which the narrator supposed to be fæcal matter, is exactly that which occurs when a cuttle discharges the contents of the remarkable organ known as its ink-bag; and the strong and peculiar scent mentioned as appertaining to it, is actually characteristic of its inky secretion. The musky odour referred to, is more perceptible in some species than in others. In one of the Octopods (*Eledone moschatus*), it is so strong, that the specific name of the animal is derived from it.

The ancient Greeks and Romans, who were well acquainted with the various kinds of cuttles and regarded them all as excellent food, and even as delicacies of the table, applied the word "polypus" especially to the octopus. But Pontoppidan evidently uses it as descriptive of all the cephalopods. It must not be forgotten, however, that when he wrote, science was only slowly recovering from neglect of many centuries' duration. In the enlightened times of Greece and Rome, natural history flourished, and as in our day, attracted and occupied the attention of the man of science, and afforded recreation to the man of business and the politician. Aristotle wrote 322 years before the birth of Christ, and his works are monuments of practical wisdom. When we consider the period during which he lived, and the isolated nature of his labours, and compare them with the information which he possessed, we are astonished at his sagacity and the great scope and general accuracy of his knowledge. Pliny, 240 years later, lived in times more favourable for the cultivation of science; but with all his advantages made little improvement on the work of the great master. And then, later still, the sun of learning set; and there came over Europe the long night of the dark ages which succeeded

Roman greatness, during which science was degraded and ignorance prevailed; and it is not till the middle of the sixteenth century, that the zoologist finds much to interest and instruct him. When we further reflect, that until within the past five and twenty years—till our large aquaria were constructed—Aristotle's knowledge of the habits and life-history of marine animals, and amongst them the cephalopods, was incomparably greater and more perfect than that possessed by any man who had lived since he recorded his observations, we cannot help feeling that in some departments of knowledge there is still lost ground to be recovered.

In the old days of the Cæsars, a Greek or Roman housewife who was accustomed to see the cuttle, the squid, and the octopus daily exposed for sale in the markets, would of course have laughed at the idea of mistaking the one for the other; but there are comparatively few persons in our own country, at the present day, except those who have made marine zoology their study, whose ideas on the subject are not exceedingly hazy. This want of technical knowledge is not confined to the masses;.but is common, if not general, amongst those who have been well educated, and is frequently apparent even in leaders in the daily papers—the productions, for the most part, of men of receptive minds, trained discrimination, and great general knowledge. As the subject is one in which I have long felt especial interest, I venture to hope that I may succeed in making clear the difference between the eight-footed octopus and its ten-footed relatives, and thus enable the reader to identify the member of the family from which we are to strip the dress and "make up" in which it masqueraded as the Kraken, and cause it to appear in its true and natural form.

One of the great primary groups or divisions of the animal kingdom is that of the soft-bodied mollusca; which includes the cuttle, the oyster, the snail, &c. It has been separated into five "classes," of which the one we have especially to notice is the *Cephalopoda*,* or "head-footed," —the animals belonging to it having their feet, or the organs which correspond with the foot of other molluscs, so attached to the head as to form a circle or coronet round the mouth. Some of these have the foot divided into eight segments, and are therefore called the *Octopoda*:† others have, in addition to the eight feet, lobes, or arms, two longer tentacular appendages, making ten in all, and are consequently called the *Decapoda*.

Of the ten-footed section of the cephalopods, there are four "families;" two only of which exist in Britain—the *Teuthidæ*, and the *Sepiidæ*. The *Teuthidæ* are the Calamaries, popularly known as "Squids," and are represented by the long-bodied *Loligo vulgaris*, that has internally along its back a gristly, translucent stiffener, shaped like a quill-pen; from which and its ink it derives its names of "calamary" (from "*calamus*," a "pen"), "pen-and-ink fish," and "sea-clerk." The *Sepiidæ* are generally known as the Cuttles proper. As a type of them we may take the common "cuttle-fish," *Sepia officinalis*, the owner of the hard, calcareous shell often thrown up on the shore, and known as "cuttle-bone," or "sea-biscuit."

It must here be remarked, that as these head-footed mollusks are not "fish," any more than lobsters, crabs, oysters, mussels, &c., which fishmongers call "shell-fish," are "fish," the word "fish" is misleading, and should be abandoned; and secondly, that the names "cuttle" and "squid," as dis-

---

* From the Greek words *cephale*, the head; and *poda*, feet.
† From *octo*, eight; and *pous* (*poda*), feet.

tinctive appellations, are unsatisfactory. The word "cuttle" is derived from "cuddle," to hug, or embrace—in allusion to the manner in which the animal seizes its prey, and enfolds it in its arms; and "squid" is derived from "squirt," in reference to its habit of squirting water or ink. But as all the known members of the class, except the pearly nautilus, *Nautilus pompilius*, have these habits in common, the distinguishing terms are hardly apposite. As, however, they are conventionally accepted and understood, I prefer to use them. As with other mollusks, so with the cephalopods, some have shells, and some are naked or have only rudimentary shells. The Argonaut, or paper nautilus, has been regarded as the analogue of the snail, which, like it, secretes an *external* shell for the protection of its soft body; and the octopus as that of the garden slug, which, having organs like those of the snail, as the octopus has organs like those of the shell-bearing argonaut, has no shell. The cuttles and squids may be compared to some of the sea-slugs, as *Aplysia* and *Bullæa*, and to some land-slugs, as *Parmacella* and *Limax*, which have an *internal* shell.

The argonaut and the other families of the cephalopods do not come within the scope of this treatise; we will therefore confine our attention to the three above mentioned. Of the anatomy and homology of the *Octopus*, *Sepia*, and *Calamary* we need say no more than will suffice to show in what manner they resemble each other, and wherein they differ, in order that we may the more clearly perceive to which of them the story of the Kraken probably owes its origin.

The octopus, the sepia, and the calamary are all constructed on one fundamental plan. A bag of fleshy muscular skin, called the mantle-sac, contains the organs of the body, heart, stomach, liver, intestines, a pair of gills by which oxygen is absorbed from the water for the purification of the blood, and an excurrent tube by which the water thus deprived of its life-sustaining gas is expelled. The outrush of water with more or less force, from this "syphon-tube," is also the principal source of locomotion when the animal is swimming, as it propels it backward—not by the striking of the expelled fluid against the surrounding water, as is generally supposed; but by the unbalanced pressure of the fluid acting inside the body in the direction in which the creature goes. Into this syphon-tube, or funnel, opens, by a special duct, the ink-bag; and from it is squirted at will the intensely black fluid therein secreted. I doubt very much the correctness of the statement mentioned by Pontoppidan and others, that the cuttle ejects its ink with a desire to lie hidden and in ambush for its intended prey, or with the intention to attract fish within its reach by their partiality for the musky odour of this secretion. It may be so, but during the long period that I had these animals under close observation at the Brighton Aquarium, I never witnessed such an incident. I believe that the emission of the ink is a symptom of fear, and is only employed as a means of concealment from a suspected enemy. I have found, that when first taken, the *Sepia*, of all its kind, is the most sensitively timid. Its keen, unwinking eye watches for and perceives the·slightest movement of its captor; and if even most cautiously looked at from above, its ink is belched forth in eddying volumes, rolling over and over like the smoke which follows the discharge of a great gun from a ship's port, and mixes with marvellous rapidity with the surrounding water. But, like all of its class, the *Sepia* is very intelligent. It soon learns to discriminate between friend and foe, and ultimately becomes very tame, and ceases to shoot its ink, unless it be teased and excited. By means of the communication between the ink-bag and the locomotor tube, it happens that when the ink is ejected, a stream of water is forcibly emitted with it, and thus the very effort for escape serves the double purpose of propelling the creature away from danger, and discolouring the water in which it moves. Oppian has well described this—

> "The endangered cuttle thus evades his fears,
> And native hoards of fluids safely wears.
> A pitchy ink peculiar glands supply
> Whose shades the sharpest beam of light defy.
> Pursued, he bids the sable fountains flow,
> And, wrapt in clouds, eludes the impending foe.
> The fish retreats unseen, while self-born night
> With pious shade befriends her parent's flight."

Professor Owen has remarked that the ejection of the ink of the cephalopods serves by its colour as a means of defence, as corresponding secretions in some of the mammalia by their odour.

It is worthy of notice that the pearly nautilus and the allied fossil forms are without this means of concealment, which their strong external shells render unnecessary for their protection.

From the sac-like body containing the various organs, protrudes a head, globose in shape, and containing a brain, and furnished with a pair of strong, horny mandibles, which bite vertically, like the beak of a parrot. By these the flesh of prey is torn and partly masticated, and within them lies the tongue, covered with recurved and retractile teeth, like that of its distant relatives, the whelk, limpet, &c., by which the food is conducted to the gullet. Around this head is, as I have said, the organ which is equivalent to the foot in other molluscs—that by which the slug and the snail crawl—only that the head is

Placed in the centre, instead of in the front of it, and it is divided into segments, which radiate from this central head. These segments are very flexible, and capable of movement in every direction, and are thus developed into arms, prehensile limbs, by which their owner can seize and hold its living prey. That this may be more perfectly accomplished, these arms are studded along their inner surface with rows of sucking discs, in each of which, by means of a retractile piston, a vacuum can be produced. The consequent pressure of the outer atmosphere or water, causes them to adhere firmly to any substance to which they are applied, whether stone, fish, crustacean, or flesh of man.

But, although in all these highly-organised head-footed mollusks the same general build prevails, it is admirably modified in each of them to suit certain habits and necessities. Thus the octopus, being a shore dweller, its soft and pliant, but very tough body, having merely a very small and rudimentary indication of an internal shell (just a little "style") is exactly adapted for wedging itself amongst crevices of rocks. A large, rigid, cellular float, or "sepiostaire," such as *Sepia* possesses, or a long, horny pen such as *Loligo* has, would be in the way, and worse than useless in such places as the octopus inhabits. Its eight long powerful arms or feet are precisely fitted for clambering over rocks and stones, and as its food of course consists principally of the living things most abundant in such localities, namely, the shore-crabs, its great flexible suckers, devoid of hooks or horny armature, are exactly adapted to firm and air-tight attachment to the smooth shells of the crustacea.

Unlike the octopus, which is capable only of short flights through the water, the "cuttles" and "squids," such as

Sepia and *Loligo*, are all free swimmers. For them it is necessary for accuracy of natation that their soft, and in the squids long bodies, should be supported by such a framework as they possess. In *Sepia*, the mantle-sac is flattened horizontally all along its lateral edges so as to form a pair of fins, which nearly surround the trunk. These fins could never be used, as they are, to enable the animal to poise itself delicately in the water by means of their beautiful undulations, which I have often watched with delight, if their attached edges were not kept in a straight line on either side. Then, these ten-footed or ten-armed genera have not, because they need them not, eight long,

FIG. I.—BEAK AND ARMS OF A DECAPOD CUTTLE.

a, the eight shorter arms ; *t*, the tentacles ; *f*, the funnel, or locomotor tube.

strong and highly mobile arms like those of the octopus, nor have they large suckers upon them. Whereas a great length of reach is an advantage to the octopus, animals which are purely swimmers, and which hunt and overtake their prey by speed, would be impeded by having to drag after them a bundle of stout, lengthy appendages trailing heavily astern. Their eight pedal arms are short and comparatively weak, though strong enough, in individuals such as are regarded on our own coasts as fullgrown, to seize and hold a fish or crustacean as strong as a good sized shore-crab. But, as compensation for the shortness of the eight arms, they are provided with two others more than three times the length of the short ones. These are so slender that they generally lie coiled up in a spiral cone in two pockets, one on each side, just below the eye, when the animal is quiescent, and are only seen when it takes its food. These long, slender tentacular arms are expanded at their extremity, and the inner surface of their enlarged part is studded with suckers—some of them larger in size than those on the eight shorter arms. As the food of these swimmers consists, of course, chiefly of fish, their sucking disks are curiously modified for the better retention of a slippery captive. A horny ring with a sharply serrated edge is imbedded in the outer circumference of each of them, and when a vacuum is formed, the keen, saw-like teeth are pressed into the skin or scales of the unfortunate prisoner, and deprive it of the slightest chance of escape.

The manner in which the eight-armed and ten-armed cephalopods capture their prey is similar in principle and plan, but differs in action in accordance with their mode of life. The ordinary habit of the octopus is either to rest suspended to the side of a rock to which it clings with the suckers of several of its arms, or to remain lurking in some favourite cranny; its body thrust for protection and concealment well back in the interior of the recess; its bright eyes keenly on the watch; three or four of its limbs firmly attached to the walls of its hiding place—the others gently waving, gliding, and feeling about in the water, as if to maintain its vigilance, and keep itself always on the alert, and in readiness to pounce on any unfortunate wayfarer that may pass near its den. To a shore-crab that comes within its reach the slightest contact with one of those lithe

FIG. 2.—THE OCTOPUS (*Octopus vulgaris*).

arms is fatal. Instantaneously as pull of trigger brings down a bird, or touch of electric wire explodes a torpedo or a mining fuse, the pistons of the series of suckers are simultaneously drawn inward, the air is removed from the pneumatic holders, and a vacuum created in each : the crab tries to escape, but in a second is completely pinioned : not a movement, not a struggle is possible ; each leg, each claw is grasped all over by suckers, enfolded in them, stretched out to its fullest extent by them ; the back of the carapace is completely covered by the tenacious disks, brought together by the adaptable contractions of the limb, and ranged in close order, shoulder to shoulder, touching each other; and the pressure of the air is so great that nothing can effect the relaxation of their retentive power but the destruction of the air-pump that works them, or the closing of the throttle-valve by which they are connected with it. Meanwhile the abdominal plates of the captive crab are dragged towards the mouth ; the black tip of the hard horny beak is seen for a single instant protruding from the circular orifice in the centre of the radiation of the arms ; and, the next, has crushed through the shell, and is buried deep in the flesh of the victim.

Unlike the skulking, hiding octopus, its ten-armed relative, the *Sepia* loves the daylight and the freedom of the upper water. Its predatory acts are not those of a concealed and ambushed brigand lying in wait behind a rock, or peeping furtively from within the gloomy shadow of a cave ; but it may better be compared to the war-like Comanche vidette seated gracefully on his horse, and scanning from some elevated knoll a wide expanse of prairie, in readiness to swoop upon a weak or unarmed foe. Poised near the surface of the water, like a hawk in the air, the *Sepia* moves gently to and fro by graceful undulations of its lateral fins,—an exquisite play of colour occasionally taking place over its beautifully barred and mottled back. When thus tranquil, its eight pedal arms are usually brought close together, and droop in front of its head, like the trunk of an elephant, shortened; its two longer tentacular arms being coiled up within their pouches and unseen. Only when some small fish approaches it does it arouse itself. Then, its eyes dilate, and its colours become more bright and vivid. It carefully takes aim, advancing or retreating to such a distance as will just allow the two hidden tentacles to reach the quarry when they shall be shot out. Next, the two highest or central feet are lifted up, and the three others on each side are spread aside, so that they may be all out of the way of the two concealed tentacles, presently to be launched forth; and then, in a moment—so instantaneously that the eye of an observer, be he ever so watchful, can hardly see the act—this pair of tentacles, side by side, are projected and withdrawn, as if in a flash. The fish or shrimp has vanished, the suckers of the dilated ends of the tentacles having adhered to it, and left it, as they re-entered their pouches, within the fatal "cuddle," or embrace, where it is torn to pieces by the devouring beak.* This action of the tentacles of the decapods is the most rapid motion that I know of in the whole animal kingdom—not excepting even that of the

---

* See an excellent article in the *Field*, Sept. 2, 1876, on the 'Ten Footed Cuttle' (*Sepia officinalis*), by the late Mr. W. A. Lloyd, an earnest and accomplished aquatic zoologist; eccentric, but in all that relates to the construction and management of an aquarium a master of his craft. It was his wish that in any future edition of my little book on the Octopus, or other writings on the cephalopods, I should use the woodcuts which illustrated his articles on Sepia and Octopus. By the kind permission of the proprietors of the *Field*, I reproduce them in suitable size for these pages.

FIG. 3.—THE CUTTLE (*Sepia officinalis*).

tongue of the toad and the lizard. These long tentacles are not used when the food is within reach of the shorter arms.

The calamaries or squids of our British Seas seize their prey in the same manner as *Sepia*, and the description of one will suffice for both. But there exist two groups of them, which are armed with curved and sharp-pointed hooks or claws, either in addition to, or instead of suckers. In the one group (*Onychoteuthis*), the hooks are restricted to the extremities of the pair of tentacles, in the other (*Enoploteuthis*), both the tentacles and the shorter arms have hooks. Professor Owen, in his description of these hook-armed calamaries in the *Cyclopædia of Anatomy*, notices also another structure which adds greatly to their prehensile power (Fig. 4.). "At the extremity of the long tentacles a cluster of small, simple, unarmed suckers may be observed at the base of the expanded part. When these latter suckers are applied to one another the tentacles are securely locked together at that part, and the united strength of both the elongated peduncles can be applied to drag towards the mouth any resisting object which has been grappled by the terminal hooks. There is no mechanical contrivance which surpasses this structure; art has remotely imitated it in the fabrication of the obstetrical forceps, in which either blade can be used separately, or, by the inter-locking of a temporary blade, be made to act in combination."

The cephalopods obtain and eat their food very much like the rapacious birds. They are the falcons of the sea. Some of them, like *Onychoteuthis*, strike their prey with talons and suckers also, others lay hold of it with suckers alone; but they all tear the flesh with their beaks, and swallow and digest their food in the same manner as the hawk or vulture.

The *Sepia*, the owner of the broad, flattened bone, has a decided predilection for the vicinity of the shore, and for comparatively shallow water. It there attaches its grape-like eggs to some convenient stone or growing alga, and delights occasionally to sink to the bottom, and there to rest half covered by the sand, a habit for which the form of its body is well adapted. But the calamaries—they of the horny pen—prefer the wide waters of the open ocean; and although they, too, especially the smaller species, are common upon the coasts, they are frequently met with far out at sea, and away from any land. The elongated and almost arrow-like shape of their bodies enables them to glide through the water with great rapidity, and the momentum exerted by a vigorous out-rush from their syphon-tube is sometimes so great that when the opposite pressure thus produced is so exerted as to cause them to take an upward direction they leap out of the water to so great a height as to fall on the decks of ships; and are, therefore, called by sailors, "flying squids." Their spawn is very different from that of either octopus, or sepia. It consists of dozens of semi-trans-

FIG. 4.—HOOKED TENTACLES OF *Onychoteuthis*.

parent, gelatinous, slender, cylindrical sheaths, about four or five inches long, each containing many ova imbedded in it (making a total number of about 40,000 embryos), all springing from a common centre and resembling a mop without a handle. I have never seen any of these "sea-mops" attached to anything, and the pelagic habits of the calamaries render it probable that they are left floating on the surface of the sea.

Having made ourselves acquainted with the structure and habits of these three divisions of the eight-footed and ten-footed mollusks, let us take evidence as to the size to which they are respectively known to attain, and the degree in which they may be regarded as dangerous to man.

An octopus from our own coasts having arms two feet in length may be considered a rather large specimen; and Dr. J. E. Gray, who was always most kindly ready to place at the disposal of any sincere inquirer the vast store of knowledge laid up in his wonderful memory, told me that "there is not one in the British Museum which exceeds this size, or which would not go into a quart pot—body, arms and all." The largest British specimen I have hitherto seen had arms 2 ft. 6 in. long. We have sufficient evidence, however, that it exceeds this in the South of France, and along the Spanish and Italian coasts of the Mediterranean; and my deceased friend John Keast Lord tells us in his book, 'The Naturalist in British Columbia,' that he saw and measured, in Vancouver's Island, an octopus which had arms five feet long.

I have often been asked whether an octopus of the ordinary size can really be dangerous to bathers. Decidedly, "Yes," in certain situations. The holding power of its numerous suckers is enormous. It is almost impossible forcibly to detach it from its adhesion to a rock or the flat bottom of a tank; and if a large one happened to fix one or more of its strong, tough arms on the leg of a swimmer whilst the others held firmly to a rock, I doubt if the man could disengage himself under water by mere strength, before being exhausted. Fortunately the octopus can be made to relax its hold by grasping it tightly round the "throat" (if I may so call it), and it may be well that this should be known.

That men are occasionally drowned by these creatures is, unhappily, a fact too well attested. I have elsewhere* related several instances of this having occurred. Omitting those, I will give two or three others which have since come under my notice. Sir Grenville Temple, in his 'Excursions in the Mediterranean Sea,' tells how a Sardinian captain, whilst bathing at Jerbeh, was seized and drowned by an octopus. When his body was found, his limbs were bound together by the arms of the animal; and this took place in water only four feet deep.

Mr. J. K. Lord's account of the formidable strength of these creatures in Oregon is confirmed by an incident recorded in the *Weekly Oregonian* (the principal paper of Oregon) of October 6th, 1877. A few days before that date an Indian woman, whilst bathing, was held beneath the surface by an octopus, and drowned. The body was discovered on the following day in the horrid embrace of the creature. Indians dived down and with their knives severed the arms of the octopus and recovered the corpse.

Mr. Clemens Laming, in his book, 'The French in Algiers,' writes :—" The soldiers were in the habit of bathing in the sea every evening, and from time to time several of them disappeared—no one knew how. Bathing was, in

* See 'The Octopus; or, the Devil-fish of Fiction and of Fact.' 1873. Chapman and Hall.

consequence, strictly forbidden; in spite of which several men went into the water one evening. Suddenly one of them screamed for help, and when several others rushed to his assistance they found that an octopus had seized him by the leg by four of its arms whilst it clung to the rock with the rest. The soldiers brought the 'monster' home with them, and out of revenge they boiled it alive and ate it. This adventure accounted for the disappearance of the other soldiers."

The Rev. W. Wyatt Gill, who for more than a quarter of a century has resided as a missionary amongst the inhabitants of the Hervey Islands, and with whom I had the pleasure of conversing on this subject when he was in England in 1875, described in the *Leisure Hour* of April 20th, 1872, another mode of attack by which an octopus might deprive a man of life. A servant of his went diving for "poulpes" (octopods), leaving his son in charge of the canoe. After a short time he rose to the surface, his arms free, but his nostrils and mouth completely covered by a large octopus. If his son had not promptly torn the living plaister from off his face he must have been suffocated—a fate which actually befel some years previously a man who foolishly went diving alone.

In *Appleton's American Journal of Science and Art*, January 31st, 1874, a correspondent describes an attack by an octopus on a diver who was at work on the wreck of a sunken steamer off the coast of Florida. The man, a powerful Irishman, was helpless in its grasp, and would have been drowned if he had not been quickly brought to the surface; for when dragged on to the raft from which he had descended, he fainted, and his companions were unable to pull the creature from its hold upon him until they had dealt it a sharp blow across its baggy body.

A similar incident occurred to the government diver of the colony of Victoria, Australia. Whilst pursuing his avocation in the estuary of the river Moyne he was seized by an octopus. He killed it by striking it with an iron bar, and brought to shore with him a portion of it with the arms more than three feet long.

Mr. Laurence Oliphant, in his 'China and Japan,' describes a Japanese show, which consisted of "a series of groups of figures carved in wood, the size of life, and as cleverly coloured as Madame Tussaud's wax-works. One of these was a group of women bathing in the sea. One of them had been caught in the folds of a cuttle-fish; the others, in alarm, were escaping, leaving their companion to her fate. The cuttle-fish was represented on a huge scale, its eyes, eyelids, and mouth being made to move simultaneously by a man inside the head."

An attack of this kind is most artistically represented in a small Japanese ivory-carving in the possession of Mr. Bartlett, of the Zoological Gardens.*

The Japanese are well acquainted with the octopus; for it is commonly depicted on their ornaments, and forms no unimportant item in their fisheries.

I have recently had an opportunity of inspecting a most curious Japanese book, in the possession of my friend Mr. W. B. Tegetmeier, which is chiefly devoted to the representations of the fisheries and fish-curing processes of the country. It is in three volumes, and is entitled, 'Land and Sea Products,' by Ki Kone. It is evidently ancient, for it is slightly worm-eaten, but the plates, each 12 inches by

* This carving was figured in illustration of an interesting paper by Professor Owen, C.B., F.R.S., &c., "On some new and rare Cephalopoda," in the Transactions of the Zoological Society, April 20, 1880.

8 inches, are full of vigour. Two of these illustrate in a very interesting manner the subject before us, and by the kindness of Mr. Tegetmeier I am able to give facsimiles of them, which appeared with an article by him on this book, in the *Field* of March 14th, 1874. Fig. 5 represents a fisherman in a boat out at sea : a gigantic octopus has thrown one of its arms over the side of the boat ; the man, who is alone, has started forward from the stern of the boat, and has succeeded, by means of a large knife attached to a long handle, in lopping off the dangerous limb of his enemy. As Mr. Tegetmeier says, " From the extreme matter of fact manner in which all these engravings are made, and the total absence of exaggeration in any other representation, I cannot but regard the relative sizes of the man, the boat, and the octopus, as correctly given, in which case we have evidence of the existence of gigantic cephalopods in Japanese waters." The only doubt I have is whether the fisherman correctly described his assailant as an octopus, and whether it was not a calamary. Fig. 6 is a vivid picture of a fishmonger's shop in a market, under the awning of which may be seen two arms of a gigantic cuttle hung up for sale as food. These are evidently of most unusual size, judging from the action of the lookers on ; the one to the left, with a tall stand or case on his back, like a Parisian cocoa-vendor, is holding out his hand in mute astonishment ; whilst the attention of the smaller personage in the right-hand corner is directed to the suspended arms of the cuttle by the man nearest to him, who is pointing to them with upraised hand. In another plate in this most interesting work a Japanese mode of fishing for cuttles is delineated. A man in a boat is tossing crabs, one at a time, into the sea, and when a cuttle rises at the bait he spears it with a trident and tosses it into the boat.

FIG. 5.—JAPANESE FISHERMAN ATTACKED BY A CUTTLE.

FIG. 6.—ARMS OF A GREAT CUTTLE EXHIBITED IN A JAPANESE FISHMONGER'S SHOP.

The octopus, therefore, though not abundant on our own coasts, is found in every sea in the temperate zone; and in so far as that it secretes an ink with which it can render the water turbid, and has many radiating arms with which it can seize and drown a man, it possesses certain attributes of the Kraken; but we have no authentic knowledge of its ever attaining to greater dimensions than I have stated, nor does it bask on the surface of the sea. It is not amongst the *Octopidæ* therefore that we must look for a solution of the mystery.

The basking condition is fulfilled by the *Sepia*; and its flattened back, supported and rendered hard and firm to the touch by the calcareous *sepiostaire* beneath the skin, is broader in proportion than that of the octopus or the squid. Thus *Sepia* might pass as a microscopic miniature of the great Scandinavian monster. But it lacks the character of size. We have no reason to believe that any true *Sepia* exists, as the family is now understood, that has a body more than eighteen inches long. If it were otherwise it would be more likely to be known of this family than of its relatives, for its lightly constructed and well known "cuttle-bone" would float on the surface for many weeks after the death of its owner, and large specimens of it would be seen and recognised from passing ships.

As we can find no species of the *Octopidæ* or *Sepiidæ* which can furnish a pretext for the stories told of the Kraken, we must try to ascertain how far a similitude to it may be traced in the third family we have discussed, the *Teuthidæ*.

The belief in the existence of gigantic cuttles is an ancient one. Aristotle mentions it, and Pliny tells of an enormous polypus which at Carteia, in Grenada—an old and important Roman colony near Gibraltar—used to come out of the sea at night, and carry off and devour salted tunnies from the curing depots on the shore; and adds that when it was at last killed, the head of it (they used to call the body the head, because in swimming it goes in advance) was found to weigh 700 lbs. Ælian records a similar incident, and describes his monster as crushing in its arms the barrels of salt fish to get at the contents. These two must have been octopods if they were anything; the word "polypus" thus especially designates it, and moreover, the free-swimming cuttles and squids would be helpless if stranded on the shore. Some of the old writers seem to have aimed rather at making their histories sensational than at carefully investigating the credibility or the contrary of the highly coloured reports brought to them. These were, of course, gross exaggerations, but there was generally a substratum of truth in them. They were based on the rare occurrence of specimens, smaller certainly, but still enormous, of some known species, and in most cases the worst that can be said of their authors is that they were culpably careless and foolishly credulous.

Unhappily so lenient a judgment cannot be passed on some comparatively recent writers. Denys de Montfort, half a century later than Pontoppidan, not only professed to believe in the Kraken, but also in the existence of another gigantic animal distinct from it; a colossal *poulpe*, or octopus, compared with which Pliny's was a mere pigmy. In a drawing fitter to decorate the outside of a showman's caravan at a fair than seriously to illustrate a work on natural history,* he depicted this tremendous cuttle as throwing its arms over a three masted vessel,

* 'Histoire Naturelle générale et particulière des Mollusques,' vol. ii., p. 256.

FIG. 7.—FACSIMILE OF DE MONTFORT'S *Poulpe colossal.*

snapping off its masts, tearing down the yards, and on the point of dragging it to the bottom, if the crew had not succeeded in cutting off its immense limbs with cutlasses and hatchets. De Montfort had good opportunities of obtaining information, for he was at one time an assistant in the geological department of the Museum of Natural History, in Paris; and wrote a work on conchology,* besides that already referred to. But it appears to have been his deliberate purpose to cajole the public; for it is reported that he exclaimed to M. Defrance: "If my entangled ship is accepted, I will make my 'colossal poulpe' overthrow a whole fleet." Accordingly we find him gravely declaring † that one of the great victories of the British navy was converted into a disaster by the monsters which are the subject of his history. He boldly asserted that the six men-of-war captured from the French by Admiral Rodney in the West Indies on the 12th of April, 1782, together with four British ships detached from his fleet to convoy the prizes, were all suddenly engulphed in the waves on the night of the battle under such circumstances as showed that the catastrophe was caused by colossal cuttles, and not by a gale or any ordinary casualty.

Unfortunately for De Montfort, the inexorable logic of facts not only annihilates his startling theory, but demonstrates the reckless falsity of his plausible statements. The captured vessels did not sink on the night of the action, but were all sent to Jamaica to refit, and arrived there safely. Five months afterwards, however, a convoy of nine line-of-battle ships (amongst which were Rodney's prizes), one frigate, and about a hundred merchantmen, were dispersed, whilst on their voyage to England, by a violent

* 'Conchyliologie Systématique.'
† 'Hist. Nat. des Moll.,' vol. ii., pp. 358 to 368.

storm, during which some of them unfortunately foundered. The various accidents which preceded the loss of these vessels was related in evidence to the Admiralty by the survivors, and official documents prove that De Montfort's fleet-destroying *poulpe* was an invention of his own, and had no part whatever in the disaster that he attributed to it.

I have been told, but cannot vouch for the truth of the report, that De Montfort's propensity to write that which was not true culminated in his committing forgery, and that he died in the galleys. But he records a statement of Captain Jean Magnus Dens, said to have been a respectable and veracious man, who, after having made several voyages to China as a master trader, retired from a seafaring life and lived at Dunkirk. He told De Montfort that in one of his voyages, whilst crossing from St. Helena to Cape Negro, he was becalmed, and took advantage of the enforced idleness of the crew to have the vessel scraped and painted. Whilst three of his men were standing on planks slung over the side, an enormous cuttle rose from the water, and threw one of its arms around two of the sailors, whom it tore away, with the scaffolding on which they stood. With another arm it seized the third man, who held on tightly to the rigging, and shouted for help. His shipmates ran to his assistance, and succeeded in rescuing him by cutting away the creature's arm with axes and knives, but he died delirious on the following night. The captain tried to save the other two sailors by killing the animal, and drove several harpoons into it; but they broke away, and the men were carried down by the monster.

The arm cut off was said to have been twenty-five feet long, and as thick as the mizen-yard, and to have had on it suckers as big as saucepan-lids. I believe the old sea-captain's narrative of the incident to be true ; the dimensions given by De Montfort are wilfully and deliberately false. The belief in the power of the cuttle to sink a ship and devour her crew is as widely spread over the surface of the globe, as it is ancient in point of time. I have been told by a friend that he saw in a shop in China a picture of a cuttle embracing a junk, apparently of about 300 tons burthen, and helping itself to the sailors, as one picks gooseberries off a bush.

Traditions of a monstrous cuttle attacking and destroying ships are current also at the present day in the Polynesian Islands. Mr. Gill, the missionary previously quoted, tells us * that the natives of Aitutaki, in the Hervey group, have a legend of a famous explorer, named Rata, who built a double canoe, decked and rigged it, and then started off in quest of adventures. At the prow was stationed the dauntless Nganaoa, armed with a long spear and ready to slay all monsters. One day when speeding pleasantly over the ocean, the voice of the ever vigilant Nganaoa was heard : " O Rata! yonder is a terrible enemy starting up from ocean depths." It proved to be an octopus (query, squid ?) of extraordinary dimensions. Its huge tentacles encircled the vessel in their embrace, threatening its instant destruction. At this critical moment Nganaoa seized his spear, and fearlessly drove it through the head of the creature. The tentacles slowly relaxed, and the dead monster floated off on the surface of the ocean.

Passing from the early records of the appearance of cuttles of unusual size, and the current as well as the traditional belief in their existence by the inhabitants of many countries, let us take the testimony of travellers and naturalists who have a right to be regarded as com-

* *Leisure Hour*, October, 1875, p. 636.

petent observers. In so doing we must bear in mind that until Professor Owen propounded the very clear and convenient classification now universally adopted, the squids, as well as the eight-footed *Octopidæ*, were all grouped under the title of *Sepia*.

Pernetty, describing a voyage made by him in the years 1763-4,* mentions gigantic cuttles met with in the Southern Seas.

Shortly afterwards, during the first week in March 1769, Banks and Solander, the scientific fellow-voyagers with Lieutenant Cook (afterwards the celebrated Captain Cook), in H.M.S. *Endeavour*, found in the North Pacific, in latitude 38° 44' S. and longitude 110° 33' W., a large calamary which had just been killed by the birds, and was floating in a mangled condition on the water. Its arms were furnished, instead of suckers, with a double row of very sharp talons, which resembled those of a cat, and, like them, were retractable into a sheath of skin from which they might be thrust at pleasure. Of this cuttle they say, with evident pleasurable remembrance of a savoury meal, they made one of the best soups they ever tasted. Professor Owen tells us, in the paper already referred to, that when he was curator of the Hunterian Museum of the Royal College of Surgeons, and preparing, in 1829, his first catalogue thereof, he was struck with the number of oceanic invertebrates which Hunter had obtained. He learned from Mr. Clift that Hunter had supplied Mr. (afterwards Sir Joseph) Banks with stoppered bottles containing alcohol, in which to preserve the new marine animals that he might meet with during the circumnavigatory voyage about to be undertaken by Cook. Thinking it probable that Banks might have stowed some parts of this great hook-armed squid in one of these bottles for

* 'Voyage aux Iles Malouines.'

his anatomical friend, he searched for, and found in a bottle marked "J. B.," portions of its arms, the beak with tongue, a heart ventricle, &c., and, amongst the dry preparations, the terminal part of the body, with an attached pair of rhomboidal fins. The remainder had furnished Cook and his companions Banks and Solander with a welcome change of diet in the commander's cabin of the *Endeavour*. As the inner surface of the arms of the squid, as well as the terminals of its tentacles, were studded with hooks, Professor Owen named it *Enoploteuthis Cookii*. He estimates the diameter of the tail fin at 15 inches, the length of its body 3 feet, of its head 10 inches, of the shorter arms 16 inches, and of the longer tentacles about the same as its body— thus giving a total length of about 6 ft. 9 in. Although individuals of other species, of larger dimensions, are known to have existed, this is the largest specimen of the hook-armed calamaries that has been scientifically examined. It would have been a formidable antagonist to a man under circumstances favourable to the exertion of its strength, and the use of its prehensile and lacerating talons.

Peron,* the well-known French zoologist, mentions having seen at sea, in 1801, not far from Van Diemen's Land, at a very little distance from his ship, *Le Géographe*, a "Sepia," of the size of a barrel, rolling with noise on the waves; its arms, between 6 and 7 feet long, and 6 or 7 inches in diameter at the base, extended on the surface, and writhing about like great snakes. He recognised in this, and no doubt correctly, one of the calamaries. The arms that he saw were evidently the animal's shorter ones, as under such circumstances, with neither enemy to combat nor prey to seize at the moment, the longer tentacles would remain concealed.

* 'Voyage de Découvertes aux Terres Australes.'

Quoy and Gaimard* report that in the Atlantic Ocean, near the Equator, they found the remains of an enormous calamary, half eaten by the sharks and birds, which could not have weighed less, when entire, than 200 lbs. A portion of this was secured, and is preserved in the Museum of Natural History, Paris.

Captain Sander Rang† records having fallen in with, in mid-ocean, a species distinct from the others, of a dark red colour, having short arms, and a body the size of a hogshead.

In a manuscript by Paulsen (referred to by Professor Steenstrup, at a meeting of Scandinavian naturalists at Copenhagen in 1847) is a description of a large calamary, cast ashore on the coast of Zeeland, which the latter named *Architeuthis monachus.* Its body measured 21 feet, and its tentacles 18 feet, making a total of 39 feet.

In 1854 another was stranded at the Skag in Jutland, which Professor Steenstrup believed to belong to the same genus as the preceding, but to be of a different species, and called it *Architeuthis dux.* The body was cut in pieces by the fishermen for bait, and furnished many wheelbarrow loads. Mr. Gwyn Jeffreys‡ says Dr. Mörch' informed him that the beak of this animal was nine inches long. He adds that another huge cephalopod was stranded in 1860 or 1861, between Hillswick and Scalloway, on the west of Shetland. From a communication received by Professor Allman, it appears that its tentacles were 16 feet long, the pedal arms about half that length, and the mantle sac 7 feet. The largest suckers examined by Professor Allman were three-quarters of an inch in diameter.

We have also the statement of the officers and crew of

* 'Voyage de l'Uranie : Zoologie,' vol. i., part 2, p. 411. 1824.
† ' Manuel des Mollusques,' p. 86.
‡ ' British Conchology,' vol. v., p. 124.

FIG. 8.—GIGANTIC CALAMARY CAUGHT BY THE FRENCH DESPATCH VESSEL 'ALECTON,' NEAR TENERIFFE.

the French despatch steamer, *Alecton*, commanded by Lieutenant Bouyer, describing their having met with a great calamary on the 30th of November, 1861, between Madeira and Teneriffe. It was seen about noon on that day floating on the surface of the water, and the vessel was stopped with a view to its capture. Many bullets were aimed at it, but they passed through its soft flesh without doing it much injury, until at length "the waves were observed to be covered with foam and blood." It had probably discharged the contents of its ink-bag; for a strong odour of musk immediately became preceptible—a perfume which I have already mentioned as appertaining to the ink of many of the cephalopoda, and also as being one of the reputed attributes of the Kraken. Harpoons were thrust into it, but would not hold in the yielding flesh; and the animal broke adrift from them, and, diving beneath the vessel, came up on the other side. The crew wished to launch a boat that they might attack it at closer quarters, but the commander forbade this, not feeling justified in risking the lives of his men. A rope with a running knot was, however, slipped over it, and held fast at the junction of the broad caudal fin; but when an attempt was made to hoist it on deck the enormous weight caused the rope to cut through the flesh, and all but the hinder part of the body fell back into the sea and disappeared. M. Berthelot, the French consul at Teneriffe, saw the fin and posterior portion of the animal on board the *Alecton* ten days afterwards, and sent a report of the occurrence to the Paris Academy of Sciences. The body of this great squid, which, like Rang's specimen, was of a deep-red colour, was estimated to have been from 16 feet to 18 feet long, without reckoning the length of its formidable arms.[*]

[*] In the accompanying illustration, the size of the squid is exaggerated, but not so much as has been supposed.

These are statements made by men who, by their intelligence, character, and position, are entitled to respect and credence ; and whose evidence would be accepted without question or hesitation in any court of law. There is, moreover, a remarkable coincidence of particulars in their several accounts, which gives great importance to their combined testimony.

But, fortunately, we are not left dependent on documentary evidence alone, nor with the option of accepting or rejecting, as caprice or prejudice may prompt us, the narratives of those who have told us they have seen what we have not. Portions of cuttles of extraordinary size are preserved in several European museums. In the collection of the Faculty of Sciences at Montpellier is one six feet long, taken by fishermen at Cette, which Professor Steenstrup has identified as *Ommastrephes pteropus*. One of the same species, which was formerly in the possession of M. Eschricht, who received it from Marseilles, may be seen in the museum at Copenhagen. The body of another, analogous to these, is exhibited in the Museum of Trieste : it was taken on the coast of Dalmatia. At the meeting of the British Association at Plymouth in 1841, Colonel Smith exhibited drawings of the beak and other parts of a very large calamary preserved at Haarlem ; and M. P. Harting, in 1860, described in the Memoirs of the Royal Scientific Academy of Amsterdam portions of two extant in other collections in Holland, one of which he believes to be Steenstrup's *Architeuthis dux*, a species which he regards as identical with *Ommastrephes todarus* of D'Orbigny.

Still there remained a residuum of doubt in the minds of naturalists and the public concerning the existence of gigantic cuttles until, towards the close of the year 1873, two specimens were encountered on the coast of Newfoundland, and a portion of one and the whole of the other, were brought ashore, and preserved for examination by competent zoologists.

The circumstances under which the first was seen, as sensationally described by the Rev. M. Harvey, Presbyterian minister of St. John's, Newfoundland, in a letter to Principal Dawson, of McGill College, were, briefly and soberly, as follows :—Two fishermen were out in a small punt on the 26th of October, 1873, near the eastern end of Belle Isle, Conception Bay, about nine miles from St. John's. Observing some object floating on the water at a short distance, they rowed towards it, supposing it to be the *débris* of a wreck. On reaching it one of them struck it with his "gaff," when immediately it showed signs of life, and shot out its two tentacular arms, as if to seize its antagonists. The other man, named Theophilus Picot, though naturally alarmed, severed both arms with an axe as they lay on the gunwale of the boat, whereupon the animal moved off, and ejected a quantity of inky fluid which darkened the surrounding water for a considerable distance. The men went home, and, as fishermen will, magnified their lost "fish." They "estimated" the body to have been 60 feet in length, and 10 feet across the tail fin; and declared that when the "fish" attacked them "it reared a parrot-like beak which was as big as a six-gallon keg."

All this, in the excitement of the moment, Mr. Harvey appears to have been willing to believe, and related without the expression of a doubt. Fortunately, he was able to obtain from the fishermen a portion of one of the tentacular arms which they had chopped off with the axe, and by so doing rendered good service to science. This fragment (Fig. 9), as measured by Mr. Alexander Murray, provincial geologist of Newfoundland, and Professor Verrill, of Yale

College, Connecticut, is 17 feet long and 3½ feet in circumference. It is now in St. John's Museum. By careful calculation of its girth, the breadth and circumference of the expanded sucker-bearing portion at its extremity, and the diameter of the suckers, Professor Verrill has computed its dimensions to have been as follows:—Length of body 10 feet; diameter of body 2 feet 5 inches. Long tentacular arms 32 feet; head 2 feet; total length about 44 feet. The upper mandible of the beak, instead of being "as large as a six-gallon keg" would be about 3 inches long, and the lower mandible 1½ inch long. From the size of the large suckers

FIG. 9.—TENTACLE OF A GREAT CALAMARY (*Architeuthis princeps*) TAKEN IN CONCEPTION BAY, NEWFOUNDLAND, OCT. 26, 1873.

relatively to those of another specimen to be presently described, he regards it as probable that this individual was a female.

In November, 1873—about three weeks after the occurrence in Conception Bay—another calamary somewhat smaller than the preceding, but of the same species, also came into Mr. Harvey's possession. Three fishermen, when hauling their herring-net in Logie Bay, about three miles from St. John's, found the huge animal entangled in its folds. With great difficulty they succeeded in despatching it and bringing it ashore, having been compelled to cut off its head before they could get it into their boat.

The body of this specimen was over 7 feet long; the caudal fin 22 inches broad; the two long tentacular arms

FIG. 10.—HEAD AND TENTACLES OF A GREAT CALAMARY (*Architeuthis princeps*) TAKEN IN LOGIE BAY, NEWFOUNDLAND, NOV. 1873.

24 feet in length; the eight shorter arms each 6 feet long, the largest of the latter being 10 inches in circumference at the base; total length of this calamary 32 feet. Professor

Verrill considers that this and the Conception Bay squid are both referable to one species—Steenstrup's *Architeuthis dux*.

Excellent woodcuts from photographs of these two specimens were given in the *Field* of December 13th, 1873, and January 31st, 1874, respectively, and I am indebted to the proprietors of that journal for their kind and courteous permission to copy them in reduced size for the illustration of this little work.

For the preservation of both of the above described specimens we have to thank Mr. Harvey, and he produces additional evidence of other gigantic cuttles having been previously seen on the coast of Newfoundland. He mentions two especially, which, as stated by the Rev. Mr. Gabriel, were cast ashore in the winter of 1870-71, near Lamaline on the south coast of the island, which measured respectively 40 feet and 47 feet in length ; and he also tells of another stranded two years later, the total length of which was 80 feet.

In the *American Journal of Science and Arts*, of March 1875, Professor Verrill gives particulars and authenticated testimony of several other examples of great calamaries, varying in total length from 30 feet to 52 feet, which have been taken in the neighbourhood of Newfoundland since the year 1870. One of these was found floating, apparently dead, near the Grand Banks in October 1871, by Captain Campbell, of the schooner *B. D. Hoskins*, of Gloucester, Mass. It was taken on board, and part of it used for bait. The body is stated to have been 15 feet long, and the pedal or shorter arms between 9 feet and 10 feet. The beak was forwarded to the Smithsonian Institution.

Another instance given by Professor Verrill is of a great squid found alive in shallow water in Coomb's Cove,

Fortune Bay, in the year 1872. Its measurements, taken by the Hon. T. R. Bennett, of English Harbour, Newfoundland, were, length of body 10 feet; length of tentacle 42 feet; length of one of the ordinary arms 6 feet: the cups on the tentacles were serrated. Professor Verrill also mentions a pair of jaws and two suckers in the Smithsonian Institution, as having been received from the Rev. A. Munn, with a statement that they were taken from a calamary which went ashore in Bonavista Bay, and which measured 32 feet in total length.

On the 22nd of September, 1877, another gigantic squid was stranded at Catalina, on the north shore of Trinity Bay, Newfoundland, during a heavy equinoctial gale. It was alive when first seen, but died soon after the ebbing of the tide, and was left high and dry upon the beach. Two fishermen took possession of it, and the whole settlement gathered to gaze in astonishment at the monster. Formerly it would have been converted into manure, or cut up as food for dogs, but, thanks to the diffusion of intelligence, there were some persons in Catalina who knew the importance of preserving such a rarity, and who advised the fishermen to take it to St. John's. After being exhibited there for two days, it was packed in half-a-ton of ice in readiness for transmission to Professor Verrill, in the hope that it would be placed in the Peabody or Smithsonian Museum; but at the last moment its owners violated their agreement, and sold it to a higher bidder. The final purchase was made for the New York Aquarium, where it arrived on the 7th of October, immersed in methylated spirit in a large glass tank. Its measurements were as follows:—length of body 10 feet; length of tentacles 30 feet; length of shorter arm 11 feet; circumference of body 7 feet; breadth of caudal fin 2 feet 9 inches; diameter of largest tentacular sucker 1 inch ; number of suckers on each of the shorter arms 250.

The appearance of so many of these great squids on the shores of Newfoundland during the term of seven years, and after so long a period of popular uncertainty as to their very existence had previously elapsed, might lead one to suppose that the waters of the North Atlantic Ocean which wash the north-eastern coasts of the American Continent were, at any rate, temporarily, their principal habitat, especially as a smaller member of their family, *Ommastrephes sagittatus*, is there found in such extraordinary numbers that it furnishes the greater part of the bait used in the Newfoundland cod fisheries. But that they are by no means confined to this locality is proved by recent instances, as well as by those already cited.

Dr. F. Hilgendorf records * observations of a huge squid exhibited for money at Yedo, Japan, in 1873, and of another of similar size, which he saw exposed for sale in the Yedo fish market.

When the French expedition was sent to the Island of St. Paul, in 1874, for the purpose of observing the transit of Venus, which occurred on the 9th of December in that year, it was fortunately accompanied by an able zoologist, M. Ch. Velain. He reports † that on the 2nd of November a tidal wave cast upon the north shore of the island a great calamary which measured in total length nearly 23 feet, namely: length of body 7 feet ; length of tentacles 16 feet. There are several points of interest connected with its generic characters, and M. Velain's grounds for regarding it as being of a previously unknown species, but they

---

* 'Sitzungsberichte der Gesellschaft naturforschenden Freunde zu Berlin,' pp. 65–67, quoted by Professor Owen, *op. cit.*

† ' Comptes Rendus,' t. 80, 1875, p. 998.

are too technical for discussion here. This specimen was photographed as it lay upon the beach by M. Cazin, the photographer to the expedition.

The following account of the still more recent capture of a large squid off the west coast of Ireland was given in the *Zoologist* of June 1875, by Sergeant Thomas O'Connor, of the Royal Irish Constabulary :—

"On the 26th of April, 1875, a very large calamary was met with on the north-west of Boffin Island, Connemara. The crew of a 'curragh' (a boat made like the 'coracle,' with wooden ribs covered with tarred canvas) observed to seaward a large floating mass, surrounded by gulls. They pulled out to it, believing it to be wreck, but to their astonishment found it was an enormous cuttle-fish, lying perfectly still, as if basking on the surface of the water. Paddling up with caution, they lopped off one of its arms. The animal immediately set out to sea, rushing through the water at a tremendous pace. The men gave chase, and, after a hard pull in their frail canvas craft, came up with it, five miles out in the open Atlantic, and severed another of its arms and the head. These portions are now in the Dublin Museum. The shorter arms measure, each, eight feet in length, and fifteen inches round the base : the tentacular arms are said to have been thirty feet long. The body sank."

Finally, there is in our own national collection, preserved in spirit in a tall glass jar, a single arm of a huge cephalopod, which, by the kindness and courtesy of the officers of the department, I was permitted to examine and measure when I first described it, in May, 1873. It is 9 feet long, and 12 inches in circumference at the base, tapering gradually to a fine point. It has about 300 suckers, pedunculated, or set on tubular footstalks, placed alternately in two rows, and having serrated, horny rings, but no hooks ; the diameter of the largest of these rings is half an inch ; the smallest is not larger than a pin's head. This is one of the eight shorter, or pedal, and not one of the long, or tentacular, arms of the calamary to which it belonged. The relative length of the arms to that of the body and tentacles varies in different genera of the *Teuthidæ*, and it is not impossible that this may be the case even in individuals of the same species. But, judging from the proportions of known examples, I estimate the length of the tentacles at 36 feet, and that of the body at from 10 to 11 feet: total length 47 feet. The beak would probably have been about 5 inches long from hinge socket to point, and the diameter of the largest suckers of the tentacles about 1 inch. So much for De Montfort's "suckers as big as saucepan-lids." From a well defined fold of skin which spreads out from each margin of that surface of the arm over which the suckers are situated, Professor Owen has given to this calamary the generic name of *Plectoteuthis*, with the specific title of *grandis* to indicate its enormous size. No history relating to this interesting specimen has been preserved. No one knows its origin, nor when it was received, but Dr. Gray told me that he believed it came from the east coast of South America. It has, however, long formed part of the stores of the British Museum, and, although previously open to public view, was more recently for many years kept in the basement chambers of the old building in Bloomsbury, which were irreverently called by the initiated "the spirit vaults and bottle department," because fishes, mollusca, &c., preserved in spirits were there deposited. I hope the public will have greater facility of access to it in the new Museum.

Here, then, in our midst, and to be seen by all who ask permission to inspect it, is, and has long been, a limb of a great cephalopod capable of upsetting a boat, or of hauling a man out of her, or of clutching one engaged in scraping a ship's side, and dragging him under water, as described by the old master-mariner Magnus Dens. The tough, supple tentacles, shot forth with lightning rapidity, would be long enough to reach him at a distance of a dozen yards, and strong enough to drag him within the grasp of the eight shorter arms, a helpless victim to the mandibles of a beak sufficiently powerful to tear him in pieces and crush some of his smaller bones. For, once within that dreadful embrace, his escape, unaided, would be impossible. The clinging power of this *Plectoteuthis* is so enormously augmented by the additional surface given by the expanded folds to the under side of the arms, that I doubt if even one of the smaller whales, such as the " White Whale," or the " Pilot Whale," could extricate itself from their combined hold, if those eight supple, clammy, adhesive arms, each 9 feet long, and 5 inches in diameter at the base on the flat under surface, and armed with a battery of 2400 suckers, were once fairly lapped around it.

Ought it to surprise us, then, that an uneducated seafaring population, such as the fishermen of Fridrichstad, mentioned by Pontoppidan, absolutely ignorant of the habits and affinities, and even unacquainted with the real external form of such a creature, should exaggerate its dimensions and invest it with mystery? All that they knew of it was that whilst their friends and neighbours, whom we will call Eric Paulsen, Hans Ohlsen, and Olaf Bruhn were out fishing one calm day, a shapeless "something" rose just above the surface of the tranquil sea not far from their boat. They could see that there was much more of its bulk under water, but how far it extended they could not ascertain. Mistrusting its appearance, and with foreboding of danger, they were about to get up their anchor, when, suddenly, from thirty feet away, a rope was shot on board which fastened itself on Hans; he was dragged from amongst them towards the strange floating mass; there was a commotion; from the foaming sea upreared themselves, as it seemed to Eric and Olaf, several writhing serpents, which twined themselves around Hans; and as they gazed, helpless, in horror and bewilderment, the monster sank, and with a mighty swirl the waters closed for ever over their unfortunate companion. The men would naturally hasten home, and describe the dreadful incident—their imagination excited by its mysterious nature; the tale would spread through the district, losing nothing by repetition, and within a week the fabled Kraken would be the result.

The existence, in almost every sea, of calamaries capable of playing their part in such a scene has been fully proved, and this vexed question of marine zoology set at rest for ever. The "much greater light on this subject," which, as Pontoppidan sagaciously foresaw, was "reserved for posterity," has been thrown upon it by the discoveries of the last few years; and the "further experience which is always the best instructor," and which he correctly anticipated would be possessed by the "future writers," to whom he bequeathed the completion of his "sketch," has been obtained. Viewed by their aid, and seen in the clearer atmosphere of our present knowledge, the great sea-monster which loomed so indefinitely vast in the mist of ignorance and superstition, stands revealed in its true form and proportions —its magnitude reduced, its outline distinct, and its mystery gone—and we recognise in the supposed Kraken, as the Norwegian bishop rightly conjectured that we should, an animal "of the Polypus (or cuttle) kind, and amongst the largest inhabitants of the ocean."

THE GREAT SEA SERPENT.

THE belief in the existence of sea-serpents of formidable dimensions is of great antiquity. Aristotle, writing about B.C. 340, says* :—"The serpents of Libya are of an enormous size. Navigators along that coast report having seen a great quantity of bones of oxen, which they believe, without doubt, to have been devoured by the serpents. These serpents pursued them when they left the shore, and upset one of their triremes"—a vessel of a large class, having three banks of oars.

Pliny tells us † that a squadron sent by Alexander the Great on a voyage of discovery, under the command of Onesicritus and Nearchus, encountered, in the neighbourhood of some islands in the Persian Gulf, sea-serpents thirty feet long, which filled the fleet with terror.

Valerius Maximus,‡ quoting Livy, describes the alarm into which, during the Punic wars, the Romans, under Attilius Regulus (who was afterwards so cruelly put to death by the Carthaginians), were thrown by an aquatic, though not marine, serpent which had its lair on the banks of the Bagrados, near Ithaca. It is said to have swallowed many of the soldiers, after crushing them in its folds, and to have kept the army from crossing the river, till at length, being invulnerable by ordinary weapons, it was destroyed by heavy stones hurled by balistas, catapults, and other military engines used in those days for casting heavy missiles, and battering the walls of

---

* 'History of Animals,' book 8, chap. 28.
† 'Naturalis Historiæ,' Lib. vi., cap. 23.
‡ 'De Factis, Dictisque Memoribilibus,' Lib. i., cap. 8, 1st century.

fortified towns. According to the historian, the annoyance caused by it to the army did not cease with its death, for the water was polluted with its gore, and the air with the noxious fumes from its corrupted carcase, to such a degree that the Romans were obliged to remove their camp. They, however secured the animal's skin and skull, which were preserved in a temple at Rome till the time of the Numantine war. This combat has been described, to the same effect, by Florus (lib. ii.), Seneca (litt. 82), Silvius Italicus (l. vi.), Aulus Gellius (lib. vi., cap. 3), Orosius, Zonaras, &c., and is referred to by Pliny (lib. viii., cap. 14) as an incident known to every one. Diodorus Siculus also tells of a great serpent, sixty feet long, which lived chiefly in the water, but landed at frequent intervals to devour the cattle in its neighbourhood. A party was collected to capture it; but their first attempt failed, and the monster killed twenty of them. It was afterwards taken in a strong net, carried alive to Alexandria, and presented to King Ptolemy II., the founder of the Alexandrian Library and Museum, who was a great collector of zoological and other curiosities. This snake was probably one of the great boas.

The "*Serpens marinus*" is figured and referred to by many other writers, but as they evidently allude to the Conger and the Murena, we will pass over their descriptions.

The sea-serpents mentioned by Aristotle, Pliny, and Diodorus were, doubtless, real sea-snakes, true marine ophidians, which are more common in tropical seas than is generally supposed. They are found most abundantly in the Indian Ocean; but they have an extensive geographical range, and between forty and fifty species of them are known. They are all highly poisonous, and some are so ferocious that they more frequently attack than avoid man.

The greatest length to which they are authentically known to attain is about twelve feet. The form and structure of these *hydrophides* are modified from those of land serpents, to suit their aquatic habits. The tail is compressed vertically, flattened from the sides, so as to form a fin like the tail of an eel, by which they propel themselves; but instead of tapering to a point, it is rounded off at the end, like the blade of a paper-knife, or the scabbard of a cavalry sabre. Like other lung-breathing animals which live in water, they are also provided with a respiratory apparatus adapted to their circumstances and requirements—their nostrils, which are very small, being furnished, like those of the seal, manatee, &c., with a valve opening at will to admit air, and closing perfectly to exclude water.

Leaving these water-snakes of the tropics, we come, next in order of date, upon some very remarkable evidence that there was current amongst a community where we should little expect to find it, the idea of a marine monster corresponding in many respects with some of the descriptions given several centuries later of the sea-serpent. In an interesting article on the Catacombs of Rome in the *Illustrated London News* of February 3rd, 1872, allusion is made by the author to the collection of sarcophagi or coffins of the early Christians, removed from the Catacombs, and preserved in the museum of the Lateran Palace, where they were arranged by the late Padre Marchi for Pope Pius IX. There are more than twenty of these, sculptured with various designs—the Father and the Son, Adam and Eve and the Serpent, the Sacrifice of Abraham, Moses striking the Rock, Daniel and the Lions, and other Scripture themes. Amongst them also is Jonah and the "whale." A facsimile of this sculpture (Fig. 11) is one of the illustrations of the article referred to. It will be seen that Jonah is being swallowed feet foremost, or possibly being ejected head first, by an enormous sea monster, having the chest and fore-legs of a horse, a long arching neck, with a mane at its base, near the shoulders, a head like nothing in nature, but having hair upon and beneath the cheeks, the hinder portion of the body being that of a serpent of prodigious length, undulating in several vertical curves. This sculpture appears to have been cut between the beginning and the middle of the third century, about

FIG. 11.—JONAH AND THE SEA MONSTER.
From the Catacombs of Rome.

A.D. 230, but it probably represents a tradition of far greater antiquity.

We will now consider the accounts given by Scandinavian historians, of the sea-serpent having been seen in northern waters. Here, I suppose, I ought to indulge in the usual flippant sneer at Bishop Pontoppidan. I know that in abstaining from doing so I am sadly out of the fashion; but I venture to think that the dead lion has been kicked at too often already, and undeservedly. Whether there be, or be not, a huge marine animal, not necessarily an ophidian, answering to some of the descriptions of the sea-serpent—so called—Pontoppidan did not invent the stories told of its appearance. Long before he was born the monster had been described and figured; and for centuries previously the Norwegians, Swedes, Danes, and Fins had believed in its existence as implicitly as in the tenets of their religious creed. Olaus Magnus, Archbishop of Upsala, in Sweden, wrote of it in A.D. 1555 as follows:*—

"They who in works of navigation on the coasts of Norway employ themselves in fishing or merchandize do all agree in this strange story, that there is a serpent there which is of a vast magnitude, namely 200 foot long, and moreover, 20 foot thick; and is wont to live in rocks and caves toward the sea-coast about Berge: which will go alone from his holes on a clear night in summer, and devour calves, lambs, and hogs, or else he goes into the sea to feed on polypus (octopus), locusts (lobsters), and all sorts of sea-crabs. He hath commonly hair hanging from his neck a cubit long, and sharp scales, and is black, and he hath flaming, shining eyes. This snake disquiets the shippers; and he puts up his head on high like a pillar, and catcheth away men, and he devours them; and this happeneth not but it signifies some wonderful change of the kingdom near at hand; namely, that the princes shall die, or be banished; or some tumultuous wars shall presently follow. There is also another serpent of an incredible magnitude in an island called Moos in the diocess of Hammer; which, as a comet portends a change in all the world, so that portends a change in the kingdom of Norway, as it was seen anno 1522; that lifts himself high above the waters, and rolls himself round like a sphere.† This serpent was thought to be fifty cubits long by conjecture, by sight afar off: there followed this the banishment of King Christiernus, and a great persecution of the Bishops; and it shewed also the destruction of the country."

The Gothic Archbishop, amongst other signs and omens, also attributes this power of divination to the small red ants which are sometimes so troublesome in houses, and declares that they also portended the downfall, A.D. 1523,

---

* 'Historia de Gentibus Septentrionalibus,' Lib. xxi. cap. 43.

† "Coils itself in spherical convolutions" is a better translation of the original Latin.

of the abominably cruel Danish king, Christian II., above mentioned. His curious work is full of wild improbabilities and odd superstitions, most of which he states with a calm air of unquestioning assent; but as he wrote in the time of our Henry VIII., long before the belief in witches and warlocks, fairies and banshees, had died out in our own country, we can hardly throw stones at him on that score. It is a most amusing and interesting history, and gives a wonderful insight of the habits and customs of the northern nations in his day.

Amongst his illustrations of the sea monsters he describes are the two of which I give facsimiles on the next page. In Fig. 12 a sea-serpent is seen writhing in many coils upon the surface of the water, and having in its mouth a sailor, whom it has seized from the deck of a ship. The poor fellow is trying to grasp the ratlins of the shrouds, but is being dragged from his hold and lifted over the bulwarks by the monster. His companions, in terror, are endeavouring to escape in various directions. One is climbing aloft by the stay, in the hope of getting out of reach in that way, whilst two others are hurrying aft to obtain the shelter of a little castle or cabin projecting over the stern. I am strongly of the opinion that this is but the fallacious representation of an actual occurrence. Read by the light of recent knowledge, these old pictures convey to a practised eye a meaning as clear as that of hieroglyphics to an Egyptologist, and my translation of this is the following: The crew of a ship have witnessed the dreadful sight of a serpent-like form issuing from the sea, rising over the bulwarks of their vessel, seizing one of their messmates from amongst them, and dragging him overboard and under water. Awe-stricken by the mysterious disappearance of their comrade, and too frightened and anxious for

FIG. 12.—A SEA SERPENT SEIZING A MAN ON BOARD SHIP.
After OLAUS MAGNUS.

FIG. 13.—A GIGANTIC LOBSTER DRAGGING A MAN FROM A SHIP.
After OLAUS MAGNUS.

their own safety to be able, during the short space of time occupied by an affair, which all happened in a few seconds, to observe accurately their terrible assailant, they naturally conjecture that it must have been a snake. It was probably a gigantic calamary, such as we now know exist, and the dead carcases of which have been found in the locality where the event depicted is supposed to have taken place. The presumed body of the serpent was one of the arms of the squid, and the two rows of suckers thereto belonging are indicated in the illustration by the medial line traversing its whole length (intended to represent a dorsal fin) and the double row of transverse septa, one on each side of it.

In Fig. 13 an enormous lobster is in the act of similarly dragging overboard from a vessel a man whom it has seized by the arm with one of its great claws. From the crude image of a lobster having eight minor claws and two larger ones, to that of a cuttle having eight minor arms and two longer ones, the transition is not great; and I believe that this also is a pictorial misrepresentation of a casualty by the attack of a calamary similar to that above described, possibly another view of the same incident. The idea is that of a sea animal capable of suddenly seizing and grasping a man, and we must remember that we have evidence, in the writings of Pontoppidan and others, that, even two centuries later than Olaus Magnus, the Norsemen's knowledge of the cuttles was exceedingly vague and indistinct. Any one who has seen, as I frequently have at the Brighton Aquarium, and as they doubtless had whilst lobster-catching, the threatening and ferocious manner in which a lobster will brandish, and, if I may use the term, "gnash" its claws at an intruding hand, even if held above the surface of the water, can well imagine a party of fishermen discussing such a tragic occurrence as the foregoing, and differing in opinion as to the identity of the creature which had caused the catastrophe, some maintaining that it must have been a sea-serpent, and others shaking their heads and asserting that nothing but a colossal lobster could have done it.

Pontoppidan, in writing his history of Norway, of course had before him the statements of Olaus Magnus ; but, though their author was an archbishop, he did not accept them with the childlike simplicity generally ascribed to him. Quoting, and, singularly enough, misquoting, the Swedish prelate as referring to a sea-serpent, when he is describing, incorrectly, one of the *Acalephæ*, or sea-nettles, Pontoppidan says :—

"I have never heard of this sort, and should hardly believe the good Olaus if he did not say that he affirmed this from his own experience. The disproportion makes me think there must be some error of the press . . . He mixes truth and fable together according to the relations of others; but this was excusable in that dark age when that author wrote. Notwithstanding all this, we, in the present more enlightened age, are much obliged to him for his industry and judicious observations."

Of the sea-serpent Pontoppidan writes :—

" I have questioned its existence myself, till that suspicion was removed by full and sufficient evidence from creditable and experienced fishermen and sailors in Norway, of which there are hundreds who can testify that they have annually seen them. All these persons agree very well in the general description ; and others who acknowledge that they only know it by report or by what their neighbours have told them, still relate the same particulars. In all my inquiry about these affairs I have hardly spoke with any intelligent person born in the manor of Nordland who was not able to give a pertinent answer, and strong assurances of the existence of this fish ; and some of our north traders that come here every year with their merchandize think it a very strange question when they are seriously asked whether there be any such creature: they think it as ridiculous as if the question was put to them whether there be such fish as eel or cod."

The worthy Bishop of Bergen did his best to sift truth from fable, but he could not always succeed in separating them. Many stupendous falsehoods were brought to him, and some of them passed through his sieve in spite of his care. Of these are the accounts of the "spawning times" of the sea-serpent, its dislike of certain scents, &c. We must pass over all this, and confine ourselves to the evidence offered by him of its having been seen.

The first witness he adduces is Captain Lawrence de Ferry, of the Norwegian navy, and first pilot in Bergen, who, premising that he had doubted a great while whether there were any such creature till he had ocular demonstration of it, made the following statement, addressed formally and officially to the procurator of Bergen:—

"Mr. JOHN REUTZ—

"The latter end of August, in the year 1746, as I was on a voyage, on my return from Trundhiem, on a very calm and hot day, having a mind to put in at Molde, it happened that when we were arrived with my vessel within six English miles of the aforesaid Molde, being at a place called Jule-Næss, as I was reading in a book, I heard a kind of a murmuring voice from amongst the men at the oars, who were eight in number, and observed that the man at the helm kept off from the land. Upon this I inquired what was the matter, and was informed that there was a sea-snake before us. I then ordered the man at the helm to keep to the land again, and to come up with this creature of which I had heard so many stories. Though the fellows were under some apprehension, they were obliged to obey my orders. In the meantime the sea-snake passed by us, and we were obliged to tack the vessel about in order to get nearer to it. As the snake swam faster than we could row, I took my gun, that was ready charged, and fired at it; on this he immediately plunged under the water. We rowed to the place where it sunk down (which in the calm might be easily observed) and lay upon our oars, thinking it would come up again to the surface; however it did not. Where the snake plunged down, the water appeared thick and red; perhaps some of the shot might wound it, the distance being very little. The head of this snake, which it held more than two feet above the surface of the water, resembled that of a horse. It was of a greyish colour, and the mouth was quite black, and very large. It had black eyes, and a long white mane, that hung down from the neck to the surface of the water. Besides the head and neck, we saw seven or eight folds, or coils, of this snake, which were very thick, and as far as we could guess there was about a fathom distance between each fold. I related this affair in a certain company, where there was a person of distinction present who desired that I would communicate to him an authentic detail of all that happened; and for this reason two of my sailors, who were present at the same time and place where I saw this monster, namely, Nicholas Pedersen Kopper, and Nicholas Nicholsen Anglewigen, shall appear in court, to declare on oath the truth of every particular herein set forth; and I desire the favour of an attested copy of the said descriptions.

"I remain, Sir, your obliged servant,

"L. DE FERRY.

"Bergen, 21st February, 1751.

"After this the before-named witnesses gave their corporal oaths, and, with their finger held up according to law, witnessed and confirmed the aforesaid letter or declaration, and every particular set forth therein to be strictly true. A copy of the said attestation was made out for the said Procurator Reutz, and granted by the Recorder. That this was transacted in our court of justice we confirm with our hand and seals. *Actum Bergis die et loco, ut supra.*

"A. C. DASS (*Chief Advocate*).
"H. C. GARTNER (*Recorder*)."

The figure of the sea-serpent (Fig. 14) given by Pontoppidan was drawn, he tells us, under the inspection of a clergyman, Mr. Hans Strom, from descriptions given of it by two of his neighbours, Messrs. Reutz and Teuchsen, of Herroe; and was declared to agree in every particular with that seen by Captain de Ferry, and another subsequently observed by Governor Benstrup. The supposed coils of the serpent's body present exactly the appearance of eight porpoises following each other in line. This is a well-known habit of some of the smaller cetacea. They are often met with at sea thus proceeding in close single file, part only of their rotund forms being visible as they raise their backs above the surface of the water to inhale air through their "blow-holes." Under these circumstances they have been described by naturalists and seamen as resembling a long string of casks or buoys, often extending for sixty, eighty, or a hundred yards. This is just such a spectacle as that described by Olaus Magnus—his "long line of spherical convolutions," and also as one reported to Pontoppidan as being descriptive of the sea-serpent:—

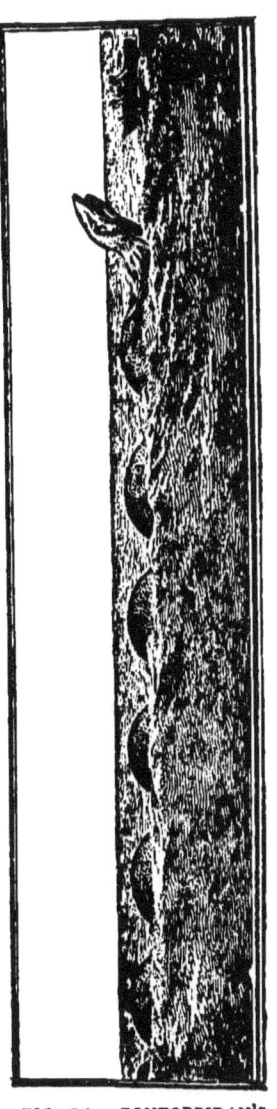

FIG. 14.—PONTOPPIDAN'S "SEA SERPENT."

"'I have been informed,' he says, ' by some of our sea-faring men that a cable * would not be long enough to measure the length of some of them when they are observed on the surface of the water in an even line. They say those round lumps or folds sometimes lie one after another as far as a man can see. I confess, if this be true, that we must suppose most probably that it is not one snake, but two or more of these creatures lying in a line that exhibit this phenomenon.' In a foot-note he adds: ' If any one enquires how many folds may be counted on a sea-snake, the answer is that the number is not always the same, but depends upon the various sizes of them: five and twenty is the greatest number that I find well attested.', Adam Olearius, in his Gottorf Museum, writes of it thus : ' A person of distinction from Sweden related here at Gottorf that he had heard the burgomaster of Malmoe, a very worthy man, say that as he was once standing on the top of a very high hill, towards the North Sea, he saw in the water, which was very calm, a snake, which appeared at that distance to be as thick as a pipe of wine, and had twenty-five folds. Those kind of snakes only appear at certain times, and in calm weather.' "

I believe that in every case so far cited from Pontoppidan, as well as that given by Olaus Magnus, the supposed coils or protuberances of the serpent's body were only so many porpoises swimming in line in accordance with their habit before mentioned. If an upraised head, like that of a horse, was seen preceding them, it was either unconnected with them, or it certainly was not that of a snake ; for no serpent could throw its body into those vertical undulations. The form of the vertebræ in the ophidians renders such a movement impossible. All their flexions are horizontal ; the curving of their body is from side to side, not up and down.

The sea-monster seen by Egede was of an entirely different kind ; and his account of it—let sceptics deride it

* Six hundred feet.

as they may—is worthy of attention and careful consideration. The Rev. Hans Egede, known as "The Apostle of Greenland," was superintendent of the Christian missions to that country. He was a truthful, pious, and single-minded man, possessing considerable powers of observation, and a genuine love of natural history. He wrote two books on the products, people, and natural history of Greenland,* and his statements therein are modest, accurate, and free from exaggeration. His illustrations are little, if at all, superior in style of art to the two Japanese wood-cuts shown on page 29, but they bear the same unmistakable signs of fidelity which characterise those of the Japanese.

In his 'Journal of the Missions to Greenland' this author tell us that—

"On the 6th of July, 1734, there appeared a very large and frightful sea monster, which raised itself so high out of the water that its head reached above our main-top. It had a long, sharp snout, and spouted water like a whale; and very broad flappers. The body seemed to be covered with scales, and the skin was uneven and wrinkled, and the lower part was formed like a snake. After some time the creature plunged backwards into the water, and then turned its tail up above the surface, a whole ship-length from the head. The following evening we had very bad weather."

The high character of the narrator would lead us to accept his statement that he had seen something previously unknown to him (he does not say it was a sea-serpent) even if we could not explain or understand what it was that he saw. Fortunately, however, the sketch made by Mr. Bing, one of his brother missionaries, has enabled us to do this. We must remember that in his endeavour to

* 'Des alten Grönlands neue perlusträtion,' 8vo., Frankfurt, 1730, and 'Det Gamle Grönlands nye perlustratione eller Naturel Historic.' 4to., Copenhagen, 1741.

FIG. 15.—THE ANIMAL DRAWN BY MR. BING AS HAVING BEEN SEEN BY HANS EGEDE.

portray the incident he was dealing with an animal with the nature of which he was unacquainted, and which was only partially, and for a very short time, within his view. He therefore delineated rather the impression left on his mind than the thing itself. But although he invested it with a character that did not belong to it, his drawing is so far correct that we are able to recognise at a glance the distorted portrait of an old acquaintance, and to say unhesi-

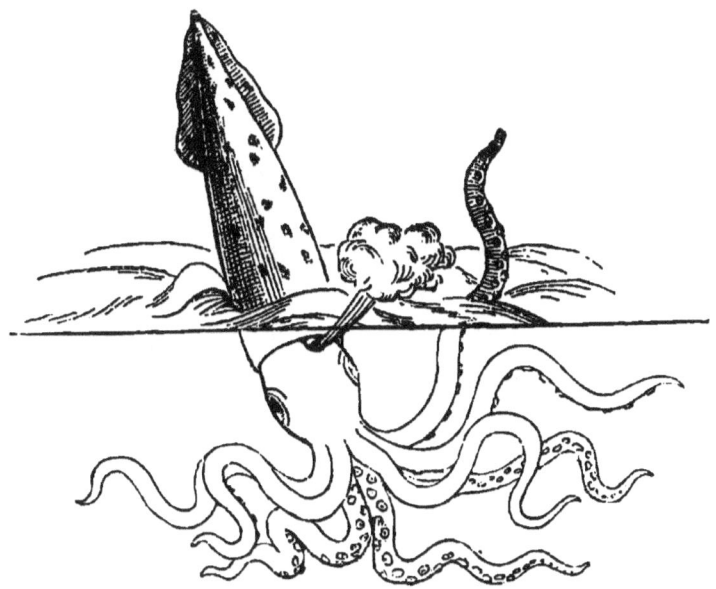

FIG. 16.—THE ANIMAL WHICH EGEDE PROBABLY SAW.

tatingly that Egede's sea-monster was one of the great calamaries which have since been occasionally met with, but which have only been believed in and recognised within the last few years. That which Mr. Egede believed to be the creature's head was the tail part of the cuttle, which goes in advance as the animal swims, and the two side appendages represent very efficiently the two lobes of the caudal fin. In propelling itself to the surface the squid raised this portion of its body out of the water to a considerable height, an occurrence which I have often witnessed, and which I have elsewhere described (see pp. 23 and 27). The supposed tail, which was turned up at some distance from the other visible portion of the body, after the latter had sunk back into the sea, was one of the shorter arms of the cuttle, and the suckers on its under side are clearly and conspicuously marked. Egede was, of course, in error in making the "spout" of water to issue from the mouth of his monster. The out-pouring jet, which he, no doubt, saw, came from the locomotor tube, and the puff of spray which would accompany it as the orifice of the tube rose to the surface of the water is sketched with remarkable truthfulness. In quoting Egede, Pontoppidan gives a copy (so-called) of this engraving, but his artist embellished it so much as to deprive it of its original force and character, and of the honestly drawn points which furnish proofs of its identity.

Pontoppidan records other supposed appearances of the sea-serpent, but from the date of his history I know of no other account of such an occurrence until that of an animal "apparently belonging to this class," which was stranded on the Island of Stronsa, one of the Orkneys, in the year 1808 :—

"According to the narrative, it was first seen entire, and measured by respectable individuals. It measured fifty-six feet in length, and twelve in circumference. The head was small, not being a foot long from the snout to the first vertebra; the neck was slender, extending to the length of fifteen feet. All the witnesses agree in assigning it blow-holes, though they differ as to the precise situation. On the shoulders something like a bristly mane commenced which extended to near the extremity of the tail. It had three pairs of fins or paws connected with the body; the anterior were the largest, measuring more than four feet in length, and their extremities were something like toes partially webbed.

The skin was smooth and of a greyish colour; the eye was of the size of a seal's. When the decaying carcass was broken up by the waves, portions of it were secured (such as the skull, the upper bones of the swimming paws, &c.) by Mr. Laing, a neighbouring proprietor, and some of the vertebræ were preserved and deposited in the Royal University Museum, Edinburgh, and in the Museum of the Royal College of Surgeons, London. An able paper," says Dr. Robert Hamilton, in his account of it,* "on these latter fragments and on the wreck of the animal was read by the late Dr. Barclay to the Wernerian Society, and will be found in Vol. I. of its Transactions, to which we refer. We have supplied a wood-cut of the sketch" (of which I give a *facsimile* here) "which was taken at the time, and which, from the many

FIG. 17.—THE "SEA SERPENT" OF THE WERNERIAN SOCIETY. (*Facsimile.*)

affidavits proffered by respectable individuals, as well as from other circumstances narrated, leaves no manner of doubt as to the existence of some such animal."

Well! one would think so. It looks convincing, and there is a savour of philosophy about it that might lull the suspicions of a doubting zoologist. What more could be required? We have accurate measurements and a sketch taken of the animal as it lay upon the shore, minute particulars of its outward form, characteristic portions of its skeleton preserved in well-known museums, and any amount of affidavits forthcoming from most respectable individuals if confirmation be required. And yet,

"'Tis true, 'tis pity;
And pity 'tis 'tis true,"

the whole fabric of circumstances crumbled at the touch

* Jardine's Naturalists' Library: 'Marine Amphibia,' p. 314.

of science. When the two vertebræ in the Museum of the Royal College of Surgeons were examined by Sir Everard Home he pronounced them to be those of a great shark of the genus *Selache*, and as being undistinguishable from those of the species called the "basking shark," of which individuals from thirty to thirty-five feet in length have been from time to time captured or stranded on our coasts. Professor Owen has confirmed this. Any one who feels inclined to dispute the identification by this distinguished comparative anatomist of a bone which he has seen and handled can examine these vertebræ for himself. If they had not been preserved, this incident would have been cited for all time as among the most satisfactorily authenticated instances on record of the appearance of the sea-serpent. As it is, it furnishes a valuable warning of the necessity for the most careful scrutiny of the evidence of well-meaning persons to whom no intentional deception or exaggeration can be imputed.

In 1809, Mr. Maclean, the minister of Eigg, in the Western Isles of Scotland, informed Dr. Neill, the secretary of the Wernerian Society, that he had seen, off the Isle of Canna, a great animal which chased his boat as he hurried ashore to escape from it; and that it was also seen by the crews of thirteen fishing-boats, who were so terrified by it that they fled from it to the nearest creek for safety. His description of it is exceedingly vague, but is strongly indicative of a great calamary.

In 1817 a large marine animal, supposed to be a serpent, was seen at Gloucester Harbour, near Cape Ann, Massachusetts, about thirty miles from Boston. The Linnæan Society of New England investigated the matter, and took much trouble to obtain evidence thereon. The depositions of eleven credible witnesses were certified on oath before magistrates, one of whom had himself seen the creature, and who confirmed the statements. All agreed that the animal had the appearance of a serpent, but estimated its length, variously, at from fifty to a hundred feet. Its head was in shape like that of a turtle, or snake, but as large as the head of a horse. There was no appearance of a mane. Its mode of progressing was by vertical undulations; and five of the witnesses described it as having the hunched protuberances mentioned by Captain de Ferry and others. Of this, I can offer no zoological explanation. The testimony given was apparently sincere, but it was received with mistrust; for, as Mr. Gosse says, "owing to a habit prevalent in the United States of supposing that there is somewhat of wit in gross exaggeration or hoaxing invention, we do naturally look with a lurking suspicion on American statements when they describe unusual or disputed phenomena."

On the 15th of May, 1833, a party of British officers, consisting of Captain Sullivan, Lieutenants Maclachlan and Malcolm of the Rifle Brigade, Lieutenant Lister of the Artillery, and Mr. Ince of the Ordnance, whilst crossing Margaret's Bay in a small yacht, on their way from Halifax to Mahone Bay, " saw, at a distance of a hundred and fifty to two hundred yards, the head and neck of some denizen of the deep, precisely like those of a common snake in the act of swimming, the head so far elevated and thrown forward by the curve of the neck, as to enable them to see the water under and beyond it. The creature rapidly passed, leaving a regular wake, from the commencement of which to the fore part, which was out of water, they judged its length to be about eighty feet." They "set down the head at about six feet in length (considerably larger than that of a horse), and that portion of the neck which they saw at the same." "There could be no mistake—no delusion," they say; "and we were all perfectly satisfied that we had been favoured with a view of the true and veritable sea-serpent." This account was published in the *Zoologist*, in 1847 (p. 1715), and at that date all the officers above named were still living.

The next incident of the kind in point of date that we find recorded carries us back to the locality of which Pontoppidan wrote, and in which was seen the animal vouched for by Captain de Ferry. In 1847 there appeared in a London daily paper a long account translated from the Norse journals of fresh appearances of the sea-serpent. The statement made was, that it had recently been frequently seen in the neighbourhood of Christiansand and Molde. In the large bight of the sea at Christiansand it had been seen every year, only in the warmest weather, and when the sea was perfectly calm, and the surface of the water unruffled. The evidence of three respectable persons was taken, namely, Nils Roe, a workman at Mr. William Knudtzon's, who saw it twice there, John Johnson, merchant, and Lars Johnöen, fisherman at Smolen. The latter said he had frequently seen it, and that one afternoon in the dog-days, as he was sitting in his boat, he saw it twice in the course of two hours, and quite close to him. It came, indeed, to within six feet of him, and, becoming alarmed, he commended his soul to God, and lay down in the boat, only holding his head high enough to enable him to observe the monster. It passed him, disappeared, and returned; but, a breeze springing up, it sank, and he saw it no more. He described it as being about six fathoms long, the body (which was as round as a serpent's) two feet across, the head as long as a ten-gallon cask, the eyes large, round, red, sparkling, and about five inches in diameter: close behind the head a mane like a fin commenced along the neck, and spread itself out on both sides, right and left, when swimming. The mane, as well as the head, was of the colour of mahogany. The body was quite smooth, its movements occasionally fast and slow. It was serpent-like, and moved up and down. The few undulations which those parts of the body and tail that were out of water made, were scarcely a fathom in length. These undulations were not so high that he could see between them and the water.

In confirmation of this account Mr. Soren Knudtzon, Dr. Hoffmann, surgeon in Molde, Rector Hammer, Mr. Kraft, curate, and several other persons, testified that they had seen in the neighbourhood of Christiansand a sea-serpent of considerable size.

Mr. William Knudtzon, and Mr. Bochlum, a candidate for holy orders, also gave their account of it, much to the same purport; but some of these remarks are worthy of note for future comment. They say, "its motions were in undulations, and so strong that white foam appeared before it, and at the side, which stretched out several fathoms. It did not appear very high out of the water; the head was long and small in proportion to the throat: as the latter appeared much greater than the former, probably it was furnished with a mane."

Sheriffe Göttsche testified to a similar effect. " He could not judge of the animal's entire length; he could not observe its extremity. At the back of the head there was a mane, which was the same colour as the rest of the body."

We must take one more Norwegian account, for it is a very important one. The venerable P. W. Deinbolt,*

* Hitherto erroneously printed "Deinboll."

Archdeacon of Molde, gives the following account of an incident that occurred there on the 28th of July, 1845:

"J. C. Lund, bookseller and printer; G. S. Krogh, merchant; Christian Flang, Lund's apprentice, and John Elgenses, labourer, were out on Romsdal-fjord, fishing. The sea was, after a warm, sunshiny day, quite calm. About seven o'clock in the afternoon, at a little distance from the shore, near the ballast place and Molde Hooe, they saw a long marine animal, which slowly moved itself forward, as it appeared to them, with the help of two fins, on the fore-part of the body nearest the head, which they judged by the boiling of the water on both sides of it. The visible part of the body appeared to be between forty and fifty feet in length, and moved in undulations, like a snake. The body was round and of a dark colour, and seemed to be several ells in thickness. As they discerned a waving motion in the water behind the animal, they concluded that part of the body was concealed under water. That it was one continuous animal they saw plainly from its movement. When the animal was about one hundred yards from the boat, they noticed tolerably correctly its fore parts, which ended in a sharp snout; its colossal head raised itself above the water in the form of a semi-circle; the lower part was not visible. The colour of the head was dark-brown and the skin smooth; they did not notice the eyes, or any mane or bristles on the throat. When the serpent came about a musket-shot near, Lund fired at it, and was certain the shots hit it in the head. After the shot it dived, but came up immediately. It raised its neck in the air, like a snake preparing to dart on his prey. After he had turned and got his body in a straight line, which he appeared to do with great difficulty, he darted like an arrow against the boat. They reached the shore, and the animal, perceiving it had come into shallow water, dived immediately and disappeared in the deep. Such is the declaration of these four men, and no one has cause to question their veracity, or imagine that they were so seized with fear that they could not observe what took place so near them. There are not many here, or on other parts of the Norwegian coast, who longer doubt the existence of the sea-serpent. The writer of this narrative was a long time sceptical, as he had not been so fortunate as to see this monster of the deep; but after the many accounts he has read, and the relations he has received from credible witnesses, he does not dare longer to doubt the existence of the sea-serpent.

"P. W. DEINBOLT.

"Molde, 29th Nov., 1845."

We may at once accept most fully and frankly the statements of all the worthy people mentioned in this series of incidents. There is no room for the shadow of a doubt that they all recounted conscientiously that which they saw. The last quoted occurrence, especially, is most accurately and intelligently described—so clearly, indeed, that it furnishes us with a clue to the identity of the strange visitant.

Here let me say—and I wish it to be distinctly understood—that I do not deny the possibility of the existence of a great sea serpent, or other great creatures at present unknown to science, and that I have no inclination to explain away that which others have seen, because I myself have not witnessed it. "Seeing is believing," it is said, and it is not agreeable to have to tell a person that, in common parlance, he "must not trust his own eyes." It seems presumptuous even to hint that one may know better what was seen than the person who saw it. And yet I am obliged to say, reluctantly and courteously, but most firmly and assuredly, that these perfectly credible eye-witnesses did not correctly interpret that which they witnessed. In these cases, it is not the eye which deceives, nor the tongue which is untruthful, but the imagination which is led astray by the association of the thing seen with an erroneous idea. I venture to say this, not with any insolent assumption of superior acumen, but because we now possess a key to the mystery which Archdeacon

Deinbolt and his neighbours had not access to, and which has only within the last few years been placed in our hands. The movements and aspect of their sea monster are those of an animal with which we are now well acquainted, but of the existence of which the narrators of these occasional visitations were unaware ; namely, the great calamary, the same which gave rise to the stories of the Kraken, and which has probably been a denizen of the Scandinavian seas and fjords from time immemorial. It must be remembered, as I have elsewhere said, that until the year 1873, notwithstanding the adventure of the *Alecton* in 1861, a cuttle measuring in total length fifty or sixty feet was generally looked upon as equally mythical with the great sea-serpent. Both were popularly scoffed at, and to express belief in either was to incur ridicule. But in the year above mentioned, specimens of even greater dimensions than those quoted were met with on the coasts of Newfoundland, and portions of them were deposited in museums, to silence the incredulous and interest zoologists. When Archdeacon Deinbolt published in 1846 the declaration of Mr. Lund and his companions of the fishing excursion, he and they knew nothing of there being such an animal. They had formed no conception of it, nor had they the instructive privilege, possessed of late years by the public in England, of being able to watch attentively, and at leisure, the habits and movements of these strangely modified mollusks living in great tanks of sea-water in aquaria. If they had been thus acquainted with them, I believe they would have recognised in their supposed snake the elongated body of a giant squid.

When swimming, these squids propel themselves backwards by the out-rush of a stream of water from a tube pointed in a direction contrary to that in which the animal is proceeding. The tail part, therefore, goes in advance, and the body tapers towards this, almost to a blunt point. At a short distance from the actual extremity two flat fins project from the body, one on each side, as shown in Figs. 16 and 18, so that this end of the squid's body somewhat resembles in shape the government "broad arrow." It is a habit of these squids, the small species of which are met with in some localities in teeming abundance, to swim on the smooth surface of the water in hot and calm weather. The arrow-headed tail is then raised out of water, to a height which in a large individual might be three feet or more; and, as it precedes the rest of the body, moving at the rate of several miles an hour, it of course looks, to a person who has never heard of an animal going tail first at such a speed, like the creature's head. The appearance of this "head" varies in accordance with the lateral fins being seen in profile or in broad expanse. The elongated, tubular-looking body gives the idea of the neck to which the "head" is attached; the eight arms trailing behind (the tentacles are always coiled away and concealed) supply the supposed mane floating on each side; the undulating motion in swimming, as the water is alternately drawn in and expelled, accords with

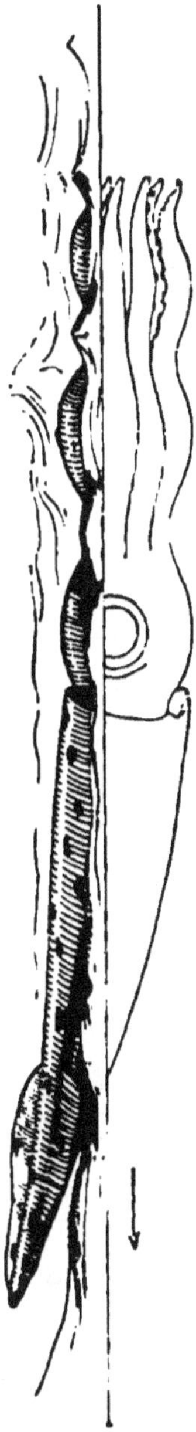

FIG. 18.—A CALAMARY SWIMMING AT THE SURFACE OF THE SEA.

the description, and the excurrent stream pouring aft from the locomotor tube, causes a long swirl and swell to be left in the animal's wake, which, as I have often seen, may easily be mistaken for an indefinite prolongation of its body. The eyes are very large and prominent, and the general tone of colour varies through every tint of brown, purple, pink, and grey, as the creature is more or less excited, and the pigmentary matter circulates with more or less vigour through the curiously moving cells.

Here we have the "long marine animal" with "two fins on the forepart of the body near the head," the "boiling of the water," the "moving in undulations," the "body round, and of a dark colour," the "waving motion in the water behind the animal, from which the witnesses concluded that part of the body was concealed under water," the "head raised, but the lower part not visible," "the sharp snout," the "smooth skin," and the appearance described by Mr. William Knudtzon, and Candidatus Theologiæ Bochlum, of "the head being long and small in proportion to the throat, the latter appearing much greater than the former," which caused them to think "it was *probably* furnished with a mane." Not that they *saw* any mane, but as they had been told of it, they thought they *ought to have seen it*. Less careful and conscientious persons would have persuaded themselves, and declared on oath, that they *did see it*.

I need scarcely point out how utterly irreconcileable is the proverbially smooth, gliding motion of a serpent, with the supposition of its passage through the water causing such frictional disturbance that "white foam appeared before it, and at the side, which stretched out several fathoms," and of "the water boiling around it on both sides of it." The cuttle is the only animal that I know of that would cause this by the effluent current from its "syphon tube." I have seen a deeply laden ship push in front of her a vast hillock of water, which fell off on each side in foam as it was parted by her bow; but that was of man's construction. Nature builds on better lines. No swimming creature has such unnecessary friction to overcome. Even the seemingly unwieldy body of a porpoise enters and passes through the water without a splash, and nothing can be more easy and graceful than the feathering action of the flippers of the awkward-looking turtle.

We now come to an incident which, from the character of those who witnessed it, immediately commanded attention, and excited popular curiosity. In the *Times* of the 9th of October, 1848, appeared a paragraph stating that a sea-serpent had been met with by the *Dædalus* frigate, on her homeward voyage from the East Indies. The Admiralty immediately inquired of her commander, Captain M'Quhæ, as to the truth of the report; and his official reply, as follows, addressed to Admiral Sir W. H. Gage, G.C.H., Devonport, was printed in the *Times* of the 13th of October, 1848.

" H.M.S. *Dædalus*, Hamoaze,
" October 11th, 1848.

" SIR,—In reply to your letter of this date, requiring information as to the truth of the statement published in the *Times* newspaper, of a sea-serpent of extraordinary dimensions having been seen from H.M.S. *Dædalus*, under my command, on her passage from the East Indies, I have the honour to acquaint you, for the information of my Lords Commissioners of the Admiralty, that at 5 o'clock P.M. on the 6th of Aug. last, in lat. 24° 44' S. and long. 9° 22' E., the weather dark and cloudy, wind fresh from the N.W. with a long ocean swell from the W., the ship on the port tack, head being N.E. by N., something very unusual was seen by Mr. Sartoris, midshipman, rapidly approaching the ship from before the beam. The circumstance was immediately reported by him to the officer of the watch, Lieut. Edgar Drummond, with whom and Mr. Wm. Barrett, the Master, I was at the time walking the quarter-deck. The ship's company were at supper. On our attention being called to the object it was discovered to be an enormous serpent, with head and shoulders kept about four feet constantly above the surface of the sea, and, as nearly as we could approximate by comparing it with the length of what our main-topsail yard would show in the water, there was at the very least sixty feet of the animal *à fleur d'eau*, no portion of which was, to our perception, used in propelling it through the water, either by vertical or horizontal undulation. It passed rapidly, but so close under our lee quarter that had it been a man of my acquaintance I should easily have recognised his features with the naked eye; and it did not, either in approaching the ship or after it had passed our wake, deviate in the slightest degree from its course to the S.W., which it held on at the pace of from twelve to fifteen miles per hour, apparently on some determined purpose.

"The diameter of the serpent was about fifteen or sixteen inches behind the head, which was without any doubt that of a snake; and it was never, during the twenty minutes it continued in sight of our glasses, once below the surface of the water; its colour dark brown, and yellowish white about the throat. It had no fins, but something like the mane of a horse, or rather a bunch of seaweed, washed about its back. It was seen by the quarter-master, the boatswain's mate, and the man at the wheel, in addition to myself and the officers above mentioned.

"I am having a drawing of the serpent made from a sketch taken immediately after it was seen, which I hope to have ready for transmission to my Lords Commissioners of the Admiralty by to-morrow's post.—PETER M'QUHÆ, Captain."

The sketches referred to in the captain's letter were made under his supervision, and copies of them, of which he certified his approbation, were published in the *Illustrated London News* on the 28th of October, 1848. I am kindly permitted by the proprietors of that journal to reproduce

THE GREAT SEA SERPENT. 81

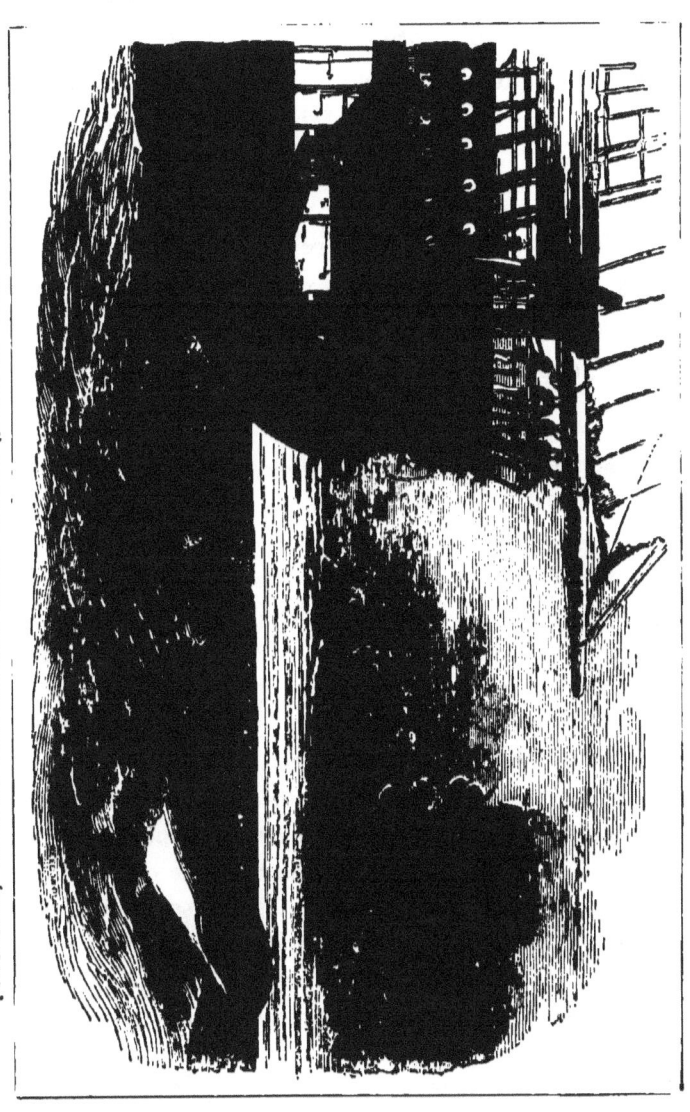

FIG. 19.—THE "SEA SERPENT" PASSING UNDER THE QUARTER OF H.M.S. 'DÆDALUS.'

G

two of them, reduced in size to suit these pages—one showing the relative positions of the "serpent" and the ship when the former was first seen (*Frontispiece*), and the other (Fig. 19) representing the animal afterwards passing under the frigate's quarter. An enlarged drawing of its head was also given, which I have not thought it necessary to copy.

Lieutenant Drummond, the officer of the watch mentioned in Captain M'Quhæ's report, published his memorandum of the impression made on his mind by the animal at the time of its appearance. It differs somewhat from the captain's description, and is the more cautious of the two.

"I beg to send you the following extract from my journal. H.M.S. 'Dædalus,' August 6, 1848, lat. 25° S., long. 9° 37' E., St. Helena 1,015 miles. In the 4 to 6 watch, at about 5 o'clock, we observed a most remarkable fish on our lee-quarter, crossing the stern in a S.W. direction. The appearance of its head, which with the back fin was the only portion of the animal visible, was long, pointed and flattened at the top, perhaps ten feet in length, the upper jaw projecting considerably; the fin was perhaps 20 feet in the rear of the head, and visible occasionally; the captain also asserted that he saw the tail, or another fin, about the same distance behind it; the upper part of the head and shoulders appeared of a dark brown colour, and beneath the under-jaw a brownish-white. It pursued a steady undeviating course, keeping its head horizontal with the surface of the water, and in rather a raised position, disappearing occasionally beneath a wave for a very brief interval, and not apparently for purposes of respiration. It was going at the rate of perhaps from twelve to fourteen miles an hour, and when nearest was perhaps one hundred yards distant; in fact it gave one quite the idea of a large snake or eel. No one in the ship has ever seen anything similar; so it is at least extraordinary. It was visible to the naked eye for five minutes, and with a glass for perhaps fifteen more. The weather was dark and squally at the time, with some sea running.—EDGAR DRUMMOND, Lieut. H.M.S. 'Dædalus;' Southampton, Oct. 28, 1848."

Statements so interesting and important, of course, elicited much correspondence and controversy. Mr. J. D. Morries Stirling, a director of the Bergen Museum, wrote to the Secretary of the British Admiralty, Captain Hamilton, R.N., saying that while becalmed in a yacht between Bergen and Sogne, in Norway, he had seen, three years previously, a large fish or reptile of cylindrical form (he would not say "sea serpent") ruffling the otherwise smooth surface of the fjord. No head was visible. This appears to have been, like the others from the same locality, a large calamary. Mr. Stirling unaware, doubtless, that Mr. Edward Newman, editor of the *Zoologist*, had previously propounded the same idea, suggested that the supposed serpent might be one of the old marine reptiles, hitherto supposed only to exist in the fossil state. This letter was published in the *Illustrated News* of October 28th, and four days afterwards, November 2nd, a letter signed F. G. S. appeared in the *Times*, in which the same idea was mooted, and the opinion expressed that it might be the *Plesiosaurus*. This brought out that great master in physiology, Professor Owen, who in a long, and, it is needless to say, most able letter to the *Times*, dated the 9th of November, 1848, set forth a series of weighty arguments against belief in the supposed serpent, which I regret that I am unable, from want of space, to quote *in extenso*. The reasoning of the most eminent of living physiologists of course had its influence on those who could best appreciate it; but, as it went against the current of popular opinion, it met with little favour from the public, and has been slurred over much too superciliously by some subsequent writers. He suggested also that the creature seen might have been a great seal, such as the leonine seal, or the sea-elephant (the head, as shown in the enlarged drawing, was wonderfully seal-like), but it was generally felt that this explanation was unsatisfactory. The nature of his criticism of the official statement will be seen from Captain M'Quhæ's reply, which was promptly given in the *Times* of the 21st of November, 1848, as follows:—

"Professor Owen correctly states that I evidently saw a large creature moving rapidly through the water very different from anything I had before witnessed, neither a whale, a grampus, a great shark, an alligator, nor any of the larger surface-swimming creatures fallen in with in ordinary voyages. I now assert—neither was it a common seal nor a sea-elephant, its great length and its totally differing physiognomy precluding the possibility of its being a '*Phoca*' of any species. The head was flat, and not a 'capacious vaulted cranium;' nor had it a stiff, inflexible trunk—a conclusion at which Professor Owen has jumped, most certainly not justified by the simple statement, that no portion of the sixty feet seen by us was used in propelling it through the water either by vertical or horizontal undulation.

"It is also assumed that the 'calculation of its length was made under a strong preconception of the nature of the beast;' another conclusion quite contrary to the fact. It was not until after the great length was developed by its nearest approach to the ship, and until after that most important point had been duly considered and debated, as well as such could be in the brief space of time allowed for so doing, that it was pronounced to be a serpent by all who saw it, and who are too well accustomed to judge of lengths and breadths of objects in the sea to mistake a real substance and an actual living body, coolly and dispassionately contemplated, at so short a distance, too, for the 'eddy caused by the action of the deeper immersed fins and tail of a rapidly moving gigantic seal raising its head above the surface of the water,' as Professor Owen imagines, in quest of its lost iceberg.

"The creative powers of the human mind may be very limited,

On this occasion they were not called into requisition; my purpose and desire throughout being to furnish eminent naturalists, such as the learned Professor, with accurate facts, and not with exaggerated representations, nor with what could by any possibility proceed from optical illusion; and I beg to assure him that old Pontoppidan having clothed his sea-serpent with a mane could not have suggested the idea of ornamenting the creature seen from the 'Dædalus' with a similar appendage, for the simple reason that I had never seen his account, or even heard of his sea-serpent, until my arrival in London. Some other solution must therefore be found for the very remarkable coincidence between us in that particular, in order to unravel the mystery.

"Finally, I deny the existence of excitement or the possibility of optical illusion. I adhere to the statements, as to form, colour, and dimensions, contained in my official report to the Admiralty, and I leave them as data whereupon the learned and scientific may exercise the 'pleasures of imagination' until some more fortunate opportunity shall occur of making a closer acquaintance with the 'great unknown'—in the present instance most assuredly no ghost.

"P. M'QUHÆ, late Captain of H.M.S. 'Dædalus.'"

Of course neither Professor Owen, nor any one else, doubted the veracity or *bona fides* of the captain and officers of one of Her Majesty's ships; and their testimony was the more important because it was that of men accustomed to the sights of the sea. Their practised eyes would, probably, be able to detect the true character of anything met with afloat, even if only partially seen, as intuitively as the Red Indian reads the signs of the forest or the trail; and therefore they were not likely to be deceived by any of the objects with which sailors are familiar. They would not be deluded by seals, porpoises, trunks of trees, or Brobdingnagian stems of algæ; but there was one animal with which they were not familiar, of the existence of which they were unaware, and which, as I have said, at that date was generally believed to be as unreal as the sea-serpent itself—namely, the great calamary, the elongated form of which has certainly in some other instances been mistaken for that of a sea-snake. One of these seen swimming in the manner I have described, and endeavoured to portray (p. 77), would fulfil the description given by Lieutenant Drummond, and would in a great measure account for the appearances reported by Captain M'Quhæ. "*The head long, pointed and flat on the top,*" accords with the pointed extremity and caudal fin of the squid. "*Head kept horizontal with the surface of the water, and in rather a raised position, disappearing occasionally beneath a wave for a very brief interval, and not apparently for purposes of respiration.*" A perfect description of the position and action of a squid swimming. "*No portion of it perceptibly used in propelling it through the water, either by vertical or horizontal undulations.*" The mode of propulsion of a squid—the outpouring stream of water from its locomotor tube—would be unseen and unsuspected, because submerged. Its effect, the swirl in its wake, would suggest a prolongation of the creature's body. The numerous arms trailing astern at the surface of the water would give the appearance of a mane. I think it not impossible that if the officers of the *Dædalus* had been acquainted with this great sea creature the impression on their mind's eye would not have taken the form of a serpent. I offer this, with much diffidence, as a suggestion arising from recent discoveries; and by no means insist on its acceptance; for Captain M'Quhæ, who had a very close view of the animal, distinctly says that "the head was, without any doubt, that of a serpent," and one of his officers subsequently declared that the eye, the mouth, the nostril, the colour, and the form were all most distinctly visible.

In a letter addressed to the Editor of the *Bombay Times*, and dated "Kamptee, January 3rd, 1849," Mr. R. Davidson, Superintending Surgeon, Nagpore Subsidiary Force, describes a great sea animal seen by him whilst on board the ship *Royal Saxon*, on a voyage to India, in 1829. The features of this incident are consistent with his having seen one of the, then unknown, great calamaries.

Dr. Scott, of Exeter, sent to the Editor of the *Zoologist* (p. 2459), an extract from the memorandum-book of Lieutenant Sandford, R.N., written about the year 1820, when he was in command of the merchant ship *Lady Combermere*. In it he mentions his having met with, in lat. 46, long. 3 (Bay of Biscay), an animal unknown to him, an immense body on the surface of the water, spouting, not unlike the blowing of a whale, and the raising up of a triangular extremity, and subsequently of a head and neck erected six feet above the surface of the water. This was evidently a great squid seen under circumstances similar to those described by Hans Egede (p. 67).

In the *Sun* Newspaper of July 9th, 1849, was published the following statement of Captain Herriman, of the ship *Brazilian*:

"On the morning of the 24th February, the ship being becalmed in lat. 26° S., long. 8° E. (about forty miles from the place where Captain M'Quhæ is said to have seen the serpent), the captain perceived something right astern, stretched along the water to a length of twenty-five or thirty feet, and perceptibly moving from the ship, with a steady sinuous motion. The head, which seemed to be lifted several feet above the water, had something resembling a mane running down to the floating portion, and within about six feet of the tail. Of course Captain Herriman, Mr. Long, his chief officer, and the passengers who saw this came to the conclusion that it must be the sea-serpent. As the 'Brazilian' was making no headway, to bring all doubts to an issue, the captain had a boat lowered, and himself standing in the bow, armed with a harpoon, approached the monster. It was found to be an immense piece of sea-weed, drifting with the current, which sets constantly to the westward in this latitude, and which, with the swell left by the subsidence of a previous gale, gave it the sinuous snake-like motion."

Captain Harrington, of the ship *Castilian*, reported in the *Times* of February 5th, 1858, that :

"On the 12th of December, 1857, N.E. end of St. Helena distant ten miles, he and his officers were startled by the sight of a huge marine animal which reared its head out of the water within twenty yards of the ship. The head was shaped like a long nun-buoy,* and they supposed it to have been seven or eight feet in diameter in the largest part, with a kind of scroll or tuft of loose skin, encircling it about two feet from the top. The water was discoloured for several hundred feet from its head, so much so that on its first appearance my impression was that the ship was in broken water."

Evidently, again, a large calamary raising its caudal extremity and fin above the surface, and discolouring the water by discharging its ink.

This was immediately followed by a letter from Captain Frederick Smith, of the ship *Pekin*, who stated that :

"On December 28th, 1848, being then in lat. 26° S., long. 6° E. (about half-way between the Cape and St. Helena), he saw a very extraordinary-looking thing in the water, of considerable length. With the telescope, he could plainly discern a huge head and neck, covered with a shaggy-looking kind of mane, which it kept lifting at intervals out of water. This was seen by all hands, and was declared to be the great sea-serpent. A boat was lowered ; a line was made fast to the 'snake,' and it was towed alongside and hoisted on board. It was a piece of gigantic sea-weed, twenty

* See illustration, p. 67.

feet long, and completely covered with snaky-looking barnacles. So like a huge living monster did this appear, that had circumstances prevented my sending a boat to it, I should certainly have believed I had seen the great sea-serpent."

In September, 1872, Mr. Frank Buckland published, in *Land and Water*, an account by the late Duke of Marlborough, of a "sea-serpent" having been seen several times within a few days, in Loch Hourn, Scotland. A sketch of it was given which almost exactly accorded with that of Pontoppidan's sea-serpent, namely, seven hunches or protuberances like so many porpoises swimming in line, preceded by a head and neck raised slightly out of water. Many other accounts have been published of the appearance of serpent-like sea monsters, but I have only space for two or three more of the most remarkable of them.

On the 10th of January, 1877, the following affidavit was made before Mr. Raffles, magistrate, at Liverpool:

"We, the undersigned officers and crew of the barque 'Pauline' (of London), of Liverpool, in the county of Lancaster, in the United Kingdom of Great Britain and Ireland, do solemnly and sincerely declare that, on July 8, 1875, in lat. 5° 13' S., long. 35° W., we observed three large sperm whales, and one of them was gripped round the body with two turns of what appeared to be a huge serpent. The head and tail appeared to have a length beyond the coils of about thirty feet, and its girth eight feet or nine feet. The serpent whirled its victim round and round for about fifteen minutes, and then suddenly dragged the whale to the bottom, head first.

"GEO. DREVAR, Master; HORATIO THOMPSON, JOHN HENDERSON LANDELLS, OWEN BAKER, and WILLIAM LEWARN.

"Again, on July 13, a similar serpent was seen, about two hundred yards off, shooting itself along the surface, head and neck being out of the water several feet. This was seen only by the captain and one ordinary seaman.

"GEORGE DREVAR, Master.

"A few moments after it was seen some 60 feet elevated perpendicularly in the air by the chief officer and the following seamen:—Horatio Thompson, Owen Baker, Wm. Lewarn. And we make this solemn declaration, conscientiously believing the same to be true."

In the *Illustrated London News*, of November 20th, 1875, there had previously appeared a letter from the Rev. E. L. Penny, Chaplain to H.M.S. *London*, at Zanzibar, describing this occurrence and also the representation of a sketch (which I am kindly permitted to reproduce here), drawn by him from the descriptions given by the captain and crew of the *Pauline*. "The whale," he said, "should have been placed deeper in the water, but he would then have been unable to depict so clearly the manner in which the animal was attacked." He adds that, "Captain Drevar is a singularly able and observant man, and those of the crew and officers with whom he conversed were singularly intelligent; nor did any of their descriptions vary from one another in the least: there were no discrepancies." The event took place whilst their vessel was on her way from Shields to Zanzibar, with a cargo of coals, for the use of H.M.S. *London*, then the guard ship on that station.

It is impossible to doubt for a moment the genuineness of the statement made by Captain Drevar and his crew, or their honest desire to describe faithfully that which they believed they had seen; but the height to which the snake is said to have upreared itself is evidently greatly exaggerated; for it is impossible that any serpent could "elevate its body some sixty feet perpendicularly in the air"—nearly one-third of the height of the Monument of the Great Fire of

THE GREAT SEA SERPENT. 91

FIG. 20.—THE "SEA SERPENT" AND SPERM WHALE AS SEEN FROM THE 'PAULINE.'

London. I have no desire to force this narrative of the master and crew of the *Pauline* into conformity with any preconceived idea. They may have seen a veritable sea-serpent; or they may have witnessed the amours of two whales, and have seen the great creatures rolling over and over that they might breathe alternately by the blow-hole of each coming to the surface of the water; or the supposed coils of the snake may have been the arms of a great calamary, cast over and around the huge cetacean. The other two appearances—1st, the animal "seen shooting itself along the surface with head and neck raised" (p. 77), and 2nd, the elevation of the body to a considerable height, as in Egede's sea monster, (p. 67), would certainly accord with this last hypothesis; but, taking the statement as it stands, it must be left for further elucidation.

On the 28th of January, 1879, a "sea-serpent" was seen from the s.s. *City of Baltimore*, in the Gulf of Aden, by Major H. W. J. Senior, of the Bengal Staff Corps. The narrator "observed a long, black object darting rapidly in and out of the water, and advancing nearer to the vessel. The shape of the head was not unlike pictures of the dragon he had often seen, with a bull-dog expression of the forehead and eyebrows. When the monster had drawn its head sufficiently out of the water, it let its body drop, as it were a log of wood, prior to darting forward under the water. This motion caused a splash of about fifteen feet in length on either side of the neck much in the 'shape of a pair of wings.'" This last particular of its appearance, as well as its movements, suggest a great calamary; but, as one with "a bull-dog expression of eyebrow, visible at 500 yards distance," does not come within my ken, I will not claim it as such.

In June 1877 Commander Pearson reported to the

Admiralty, that on the 2nd of that month, he and other officers of the Royal Yacht *Osborne*, had seen, off Cape Vito, Sicily, a large marine animal, of which the following account and sketches were furnished by Lieutenant Haynes,

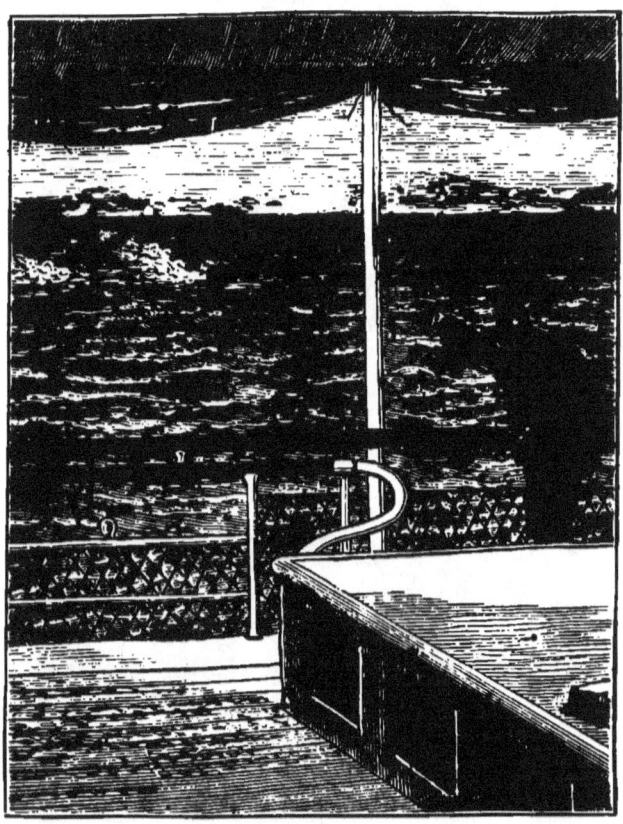

FIG. 21.—THE "SEA SERPENT" AS SEEN FROM THE 'CITY OF BALTIMORE.'

and were confirmed by Commander Pearson, Mr. Douglas Haynes, Mr. Forsyth, and Mr. Moore, engineer.

" Lieutenant Haynes writes, under date, ' Royal Yacht *Osborne*, Gibraltar, June 6': On the evening of that day, the sea being perfectly smooth, my attention was first called by seeing a ridge of fins above the surface of the water, extending about thirty feet, and varying from five to six feet in height. On inspecting it by means of a

FIG. 22.—THE "SEA SERPENT" AS SEEN FROM H.M. YACHT 'OSBORNE.'
PHASE 1.

telescope, at about one and a-half cables' distance, I distinctly saw a head, two flappers, and about thirty feet of an animal's shoulder.

FIG. 23.—THE "SEA SERPENT" AS SEEN FROM H.M. YACHT 'OSBORNE.'
PHASE 2.

The head, as nearly as I could judge, was about six feet thick, the neck narrower, about four to five feet, the shoulder about fifteen feet across, and the flappers each about fifteen feet in length. The movements of the flappers were those of a turtle, and the animal resembled a huge seal, the resemblance being strongest about the back of the head. I could not see the length of the head, but from its crown or top to just below the shoulder (where it became immersed), I should reckon about fifty feet. The tail end I did not see, being under water, unless the ridge of fins to which my attention was first attracted, and which had disappeared by the time I got a telescope, were really the continuation of the shoulder to the end of the object's body. The animal's head was not always above water, but was thrown upwards, remaining above for a few seconds at a time, and then disappearing. There was an entire absence of 'blowing,' or 'spouting.' I herewith beg to enclose a rough sketch, showing the view of the 'ridge of fins,' and also of the animal in the act of propelling itself by its two fins."

It seems to me that this description cannot be explained as applicable to any one animal yet known. The ridge of dorsal fins might, possibly, as was suggested by Mr. Frank Buckland, belong to four basking sharks, swimming in line, in close order; but the combination of them with long flippers, and the turtle-like mode of swimming, forms a zoological enigma which I am unable to solve.

This brings us face to face with the question: "Is it then so impossible that there may exist some great sea creature, or creatures, with which zoologists are hitherto unacquainted, that it is necessary in every case to regard the authors of such narratives as wilfully untruthful, or mistaken in their observations, if their descriptions are irreconcileable with something already known?" I, for one, am of the opinion that there is no such impossibility. Calamaries or squids of the ordinary size have, from time immemorial, been amongst the commonest and best known of marine animals in many seas; but only a few years ago any one who expressed his belief in one formidable enough to capsize a boat, or pull a man out of one, was derided for his credulity, although voyagers had constantly reported that in the Indian seas they were so dreaded that the natives always carried hatchets with them in their canoes, with which to cut off the arms or tentacles of these creatures, if attacked by them. We now know that their existence is no fiction; for individuals have been captured measuring more than fifty feet, and some are reported to have measured eighty feet, in total length. As marine snakes some feet in length, and having fin-like tails adapted for swimming, abound over an extensive geographical range, and are frequently met with far at sea, I cannot regard it as impossible that some of these also may attain to an abnormal and colossal development. Dr. Andrew Wilson, who has given much attention to this subject, is of the opinion that "in this huge development of ordinary forms we discover the true and natural law of the production of the giant serpent of the sea." It goes far, at any rate, towards accounting for its supposed appearance. I am convinced that, whilst naturalists have been searching amongst the vertebrata for a solution of the problem, the great unknown, and therefore unrecognized, calamaries by their elongated, cylindrical bodies and peculiar mode of swimming, have played the part of the sea-serpent in many a well-authenticated incident. In other cases, such as some of those mentioned by Pontoppidan, the supposed "vertical undulations" of the snake seen out of water have been the burly bodies of so many porpoises swimming in line—the connecting undulations beneath the surface have been supplied by the imagination. The dorsal fins of basking sharks, as figured by Mr. Buckland, or of ribbon-fishes, as suggested by Dr. Andrew Wilson, may have furnished the "ridge of fins;" an enormous conger is not an impossibility; a giant turtle may have done duty, with its propelling flippers and broad back; or a marine snake of enormous size may, really, have been seen. But if we accept as accurate the observations recorded (which I certainly do not in all cases, for they are full of errors and mistakes), the difficulty is not entirely met, even by this last admission, for the instances are very few in which an ophidian proper—a true serpent—is indicated. There has seemed to be wanting an animal having a long snake-like neck, a small head and a slender body, and propelling itself by paddles.*

The similarity of such an animal to the *Plesiosaurus* of old was remarkable. That curious compound reptile, which has been compared with "a snake threaded through the body of a turtle," is described by Dean Buckland, in his *Bridgewater Treatise*, as having "the head of a lizard, the teeth of a crocodile, a neck of enormous length resembling the body of a serpent, the ribs of a chameleon, and the paddles of a whale." In the number of its cervical vertebræ (about thirty-three) it surpasses that of the longest-necked bird, the swan.

The form and probable movements of this ancient saurian agree so markedly with some of the accounts given of the "great sea-serpent," that Mr. Edward Newman advanced the opinion that the closest affinities of the latter would be found to be with the *Enaliosauria*, or marine lizards, whose

---

* It must be noted, however, that in almost every case, except that of the *Osborne*, the paddles were *supposed*, not *seen*, and were invented to account for an animal of great length progressing at the surface of the water at the rate of twelve to fifteen miles an hour without its being possible to perceive, upon the closest and most attentive inspection, any undulatory movement to which its rapid advance could be ascribed. As the great calamaries were unknown, their mode of swift retrograde motion, by means of an outflowing current of water, was of course unsuspected.

fossil remains are so abundant in the oolite and the lias. This view has also been taken by other writers, and emphatically by Mr. Gosse. Neither he nor Mr. Newman insist that the "great unknown" must be the *Plesiosaurus* itself. Mr. Gosse says, "I should not look for any species, scarcely even any genus, to be perpetuated from the oolitic period to the present. Admitting the actual continuation of the order *Enaliosauria*, it would be, I think, quite in conformity with general analogy to find some salient features of several extinct forms."

The form and habits of the recently-recognized gigantic cuttles account for so many appearances which, without knowledge of them, were inexplicable when Mr. Gosse and Mr. Newman wrote, that I think this theory is not now forced upon us. Mr. Gosse well and clearly sums up the evidence as follows: "Carefully comparing the independent narratives of English witnesses of known character and position, most of them being officers under the crown, we have a creature possessing the following characteristics: 1st. The general form of a serpent. 2nd. Great length, say above sixty feet. 3rd. Head considered to resemble that of a serpent. 4th. Neck from twelve to sixteen inches in diameter. 5th. Appendages on the head,

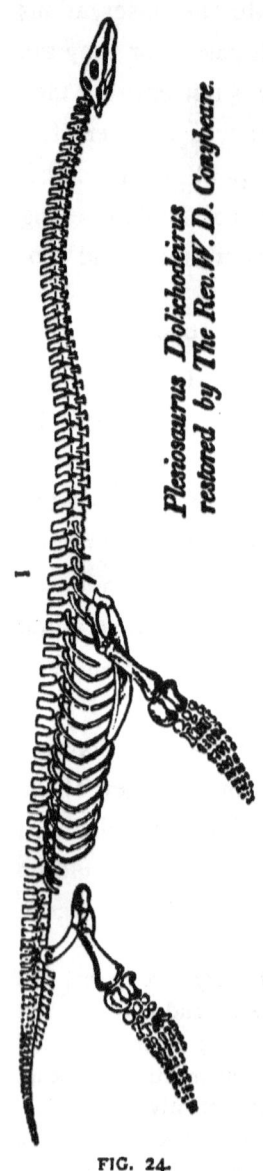

FIG. 24.

Plesiosaurus Dolichodeirus restored by The Rev. W. D. Conybeare.

neck, or back, resembling a crest or mane. (Considerable discrepancy in details.) 6th. Colour dark brown, or green, streaked or spotted with white. 7th. Swims at surface of the water with a rapid or slow movement, the head and neck projected and elevated above the surface. 8th. Progression, steady and uniform; the body straight, but capable of being thrown into convolutions. 9th. Spouts in the manner of a whale. 10th. Like a long nun-buoy." He concludes with the question—"To which of the recognized classes of created beings can this huge rover of the ocean be referred?"

I reply: "To the Cephalopoda. There is not one of the above judiciously summarized characteristics that is not supplied by the great calamary, and its ascertained habits and peculiar mode of locomotion.

Only a geologist can fully appreciate how enormously the balance of probability is contrary to the supposition that any of the gigantic marine saurians of the secondary deposits should have continued to live up to the present time. And yet I am bound to say, that this does not amount to an impossibility, for the evidence against it is entirely negative. Nor is the conjecture that there may be in existence some congeners of these great reptiles inconsistent with zoological science. Dr. J. E. Gray, late of the British Museum, a strict zoologist, is cited by Mr. Gosse as having long ago expressed his opinion that some undescribed form exists which is intermediate between the tortoises and the serpents.*

* Dr. Gray wrote in his 'Synopsis of Genera of Reptiles,' in the Annals of Philosophy, 1825: "There is every reason to believe from general structure that there exists an affinity between the tortoises and the snakes; but the genus that exactly unites them is at present unknown to European naturalists; which is not astonishing when we consider the immense number of undescribed animals which are daily

FIG. 25.—THE "SEA SERPENT," ON THE ENALIOSAURIAN HYPOTHESIS.

Professor Agassiz, too, is adduced by a correspondent of the *Zoologist* (p. 2395), as having said concerning the present existence of the *Enaliosaurian* type that "it would be in precise conformity with analogy that such an animal should exist in the American Seas, as he had found numerous instances in which the fossil forms of the Old World were represented by living types in the New."

On this point, Mr. Newman records, in the *Zoologist* (p. 2356), an actual testimony which he considers, "in all respects, the most interesting natural-history fact of the present century." He writes :

"Captain the Hon. George Hope states that when in H.M.S. 'Fly,' in the Gulf of California, the sea being perfectly calm and transparent, he saw at the bottom a large marine animal with the head and general figure of the alligator, except that the neck was much longer, and that instead of legs the creature had four large flappers, somewhat like those of turtles, the anterior pair being larger than the posterior; the creature was distinctly visible, and all its movements could be observed with ease; it appeared to be pursuing its prey at the bottom of the sea; its movements were somewhat serpentine, and an appearance of annulations, or ring-like divisions of the body, was distinctly perceptible. Captain Hope made this relation in company, and as a matter of conversation. When I heard it from the gentleman to whom it was narrated, I enquired whether Captain Hope was acquainted with those remarkable fossil animals *Ichthyosauri* and *Plesiosauri*, the supposed forms of which so nearly correspond with what he describes as having seen alive, and I cannot find that he had heard of them; the alligator being the only animal he mentioned as bearing a partial similarity to the creature in question."

---

occurring. If I may be allowed to speculate from the peculiarities of structure which I have observed, I am inclined to think that the union will most probably take place by some newly discovered genera allied to the marine or fluviatile soft-skinned turtles and the marine serpent."

Unfortunately, the estimated dimensions of this creature are not given.

That negative evidence alone is an unsafe basis for argument against the existence of unknown animals, the following illustrations will show:

During the deep-sea dredgings of H.M.S. *Lightning*, *Porcupine*, and *Challenger*, many new species of mollusca, and others which had been supposed to have been extinct ever since the chalk epoch, were brought to light; and by the deep-sea trawlings of the last-mentioned ship, there have been brought up from great depths fishes of unknown species, and which could not exist near the surface, owing to the distension and rupture of their air-bladder when removed from the pressure of deep water.

Mr. Gosse mentions that the ship in which he made the voyage to Jamaica was surrounded in the North Atlantic, for seventeen continuous hours by a troop of whales of large size of an undescribed species, which on no other occasion has fallen under scientific observation. Unique specimens of other cetaceans are also recorded.

We have evidence, to which attention has been directed by Mr. A. D. Bartlett, that "even on land there exists at least one of the largest mammals, probably in thousands, of which only one individual has been brought to notice, namely, the hairy-eared, two horned rhinoceros (*R. lasiotis*), now in the Zoological Gardens, London. It was captured in 1868, at Chittagong, in India, where for years collectors and naturalists have worked and published lists of the animals met with, and yet no knowledge of this great beast was ever before obtained, nor is there any portion of one in any museum. It remains unique.

I arrive, then, at the following conclusions: 1st. That, without straining resemblances, or casting a doubt upon narratives not proved to be erroneous, the various appearances of the supposed "Great Sea-serpent" may now be nearly all accounted for by the forms and habits of known animals; especially if we admit, as proposed by Dr. Andrew Wilson, that some of them, including the marine snakes, may, like the cuttles, attain to an extraordinary size.

2nd. That to assume that naturalists have perfect cognizance of every existing marine animal of large size, would be quite unwarrantable. It appears to me more than probable that many marine animals, unknown to science, and some of them of gigantic size, may have their ordinary habitat in the great depths of the sea, and only occasionally come to the surface; and I think it not impossible that amongst them may be marine snakes of greater dimensions than we are aware of, and even a creature having close affinities with the old sea-reptiles whose fossil skeletons tell of their magnitude and abundance in past ages.

It is most desirable that every supposed appearance of the "Great Sea-serpent" shall be faithfully noted and described; and I hope that no truthful observer will be deterred from reporting such an occurrence by fear of the disbelief of naturalists, or the ridicule of witlings.

FINIS.

LONDON:
PRINTED BY WILLIAM CLOWES AND SONS, LIMITED,
STAMFORD STREET AND CHARING CROSS.

A MERMAID.
From a Picture by Otto Sinding.

International Fisheries Exhibition
LONDON, 1883

SEA FABLES EXPLAINED

BY

HENRY LEE, F.L.S., F.G.S., F.Z.S.

SOMETIME NATURALIST OF THE BRIGHTON AQUARIUM

AND

AUTHOR OF 'THE OCTOPUS, OR THE DEVIL-FISH OF FICTION AND FACT;'
'SEA MONSTERS UNMASKED,' ETC.

ILLUSTRATED

LONDON
WILLIAM CLOWES AND SONS, Limited
INTERNATIONAL FISHERIES EXHIBITION
AND 13 CHARING CROSS, S.W.
1883

LONDON:
PRINTED BY WILLIAM CLOWES AND SONS, LIMITED,
STAMFORD STREET AND CHARING CROSS.

PREFACE.

THE little book 'Sea Monsters Unmasked,' recently issued as one of the Handbooks in connection with the Great International Fisheries Exhibition has met with so favourable a reception, that I have been honoured by the request to continue the subject, and to treat also of some of the Fables of the Sea, which once were universally believed, and even now are not utterly extinct.

The topic is not here exhausted. Other sea fables and fallacies might be mentioned and explained; but the amount of letter-press, and the number of illustrations that can be printed without loss for the small sum of one shilling—the price at which these Handbooks are uniformly published—is necessarily limited. I have, therefore, thought it better to endeavour to make each chapter as complete as possible than to crowd into the space allotted to me a greater variety of subjects less fully and carefully discussed.

I have the pleasure of acknowledging the kind assistance I have again received in the matter of illustrations. I gratefully appreciate Mr. Murray's permission to use the woodcut of Hercules slaying the Hydra, taken from Smith's 'Classical Dictionary,' and those of the golden ornaments found by Dr. Schliemann at Mycenæ, and figured in the very interesting book in which his excavations there are described. I have also to thank the proprietors of the *Illustrated London News*, the *Leisure Hour*, and *Land and Water*, for the use of illustrations especially mentioned in the text.

HENRY LEE.

SAVAGE CLUB;
Sept. 4th, 1883.

CONTENTS.

	PAGE
THE MERMAID	1
THE LERNEAN HYDRA	48
SCYLLA AND CHARYBDIS	59
THE "SPOUTING" OF WHALES	62
THE "SAILING" OF THE NAUTILUS	76
BARNACLE GEESE—GOOSE BARNACLES	98

LIST OF ILLUSTRATIONS.

FIG.		PAGE
	A MERMAID. *From a picture by Otto Sinding* . .	*Frontispiece*
1.	NOAH, HIS WIFE AND THREE SONS, AS FISH-TAILED DEITIES. *From a gem in the Florentine Gallery. After Calmet* . .	2
2.	HEA, OR NOAH, THE GOD OF THE FLOOD. *Khorsabad* . .	3
3.	DAGON. *From a bas-relief. Nimroud*	4
4.	DAGON: HALF MAN, HALF FISH. *From Lamy's 'Apparatus Biblicus'*	5
5.	DAGON. *From an agate signet. Nineveh*	,,
6.	FISH AVATAR OF VISHNU. *After Calmet and Maurice* . .	6
7.	ATERGATIS, THE GODDESS OF THE SYRIANS. *From a Phœnician Coin.*	8
8.	VENUS RISING FROM THE SEA, SUPPORTED BY TRITONS. *After Calmet*	9
9.	VENUS DRAWN IN HER CHARIOT BY TRITONS. *From two Corinthian Coins*	10
10.	DITTO.	11
11.	SEAL, DRAWN AS A FISH. *From the Catacombs at Rome* . .	,,
12.	MERMAID AND FISHES OF AMBOYNA. *After Valentyn* . .	17
13.	A JAPANESE ARTIFICIAL MERMAID	27
14.	AN ARTIFICIAL MERMAID. *Probably Japanese* . . .	28
15.	PORTRAIT OF A MERMAID SAID TO HAVE BEEN CAPTURED IN JAPAN	29
16.	THE DUGONG. *From Sir J. Emerson Tennent's 'Ceylon'* . .	43
17.	THE MANATEE	45
18.	FIGURE OF A CALAMARY, FROM THE TEMPLE OF BAYR-EL-BAHREE.	50
19.	FIGURE OF AN OCTOPUS ON A GOLD ORNAMENT FOUND BY DR. SCHLIEMANN AT MYCENÆ	51
20.	DITTO.	52
21.	DITTO.	53
22.	DITTO.	,,
23.	HERCULES SLAYING THE LERNEAN HYDRA	57

b

LIST OF ILLUSTRATIONS.

FIG.		PAGE
24.	THE PHYSETER INUNDATING A SHIP. *After Olaus Magnus*	64
25.	A WHALE POURING WATER INTO A SHIP FROM ITS BLOW-HOLE. *After Olaus Magnus*	64
26.	SPERM WHALES "SPOUTING".	65
27.	THE PAPER NAUTILUS (*Argonauta argo*) SAILING	76
28.	DITTO. RETRACTED WITHIN ITS SHELL.	81
29.	DITTO. CRAWLING.	86
30.	DITTO. SWIMMING.	87
31.	SHELL OF THE PAPER NAUTILUS (*Argonauta argo*)	88
32.	SHELL OF THE PEARLY NAUTILUS (*Nautilus pompilius*)	89
33.	THE PEARLY NAUTILUS (*Nautilus pompilius*) AND SECTION OF ITS SHELL	90
34.	THE GOOSE-TREE. *From Gerard's 'Herball'*	104
35.	DITTO. *Fac-simile from Aldrovandus*	110
36.	DEVELOPMENT OF BARNACLES INTO GEESE. *Fac-simile from Aldrovandus*	111
37.	SECTION OF A SESSILE BARNACLE. *Balanus tintinnabulum*	113
38.	PEDUNCULATED BARNACLE. *Lepas anatifera*	115
39.	A SHIP'S FIGURE-HEAD PARTLY COVERED WITH BARNACLES	116
40.	WHALE BARNACLE. *Coronula diadema*	117
41.	A YOUNG BARNACLE. *Larva of Chthamalus stellatus*	118

SEA FABLES EXPLAINED.

THE MERMAID.

NEXT to the pleasure which the earnest zoologist derives from study of the habits and structure of living animals, and his intelligent appreciation of their perfect adaptation to their modes of life, and the circumstances in which they are placed, is the interest he feels in eliminating fiction from truth, whilst comparing the fancies of the past with the facts of the present. As his knowledge increases, he learns that the descriptions by ancient writers of so-called "fabulous creatures" are rather distorted portraits than invented falsehoods, and that there is hardly one of the monsters of old which has not its prototype in Nature at the present day. The idea of the Lernean Hydra, whose heads grew again when cut off by Hercules, originated, as I have shown in another chapter, in a knowledge of the octopus; and in the form and movements of other animals with which we are now familiar we may, in like manner, recognise the similitude and archetype of the mermaid.

But we must search deeply into the history of mankind to discover the real source of a belief that has prevailed in almost all ages, and in all parts of the world, in the existence of a race of beings uniting the form of man with that of the fish. A rude resemblance between these creatures of imagination and tradition and certain aquatic animals is not sufficient to account for that belief. It probably had its origin in ancient mythologies, and in the sculptures and pictures connected with them, which were designed to represent certain attributes of the deities of various nations. In the course of time the meaning of these was lost; and subsequent generations regarded as

FIG. 1.—NOAH, HIS WIFE, AND THREE SONS, AS FISH-TAILED DEITIES.
From a Gem in the Florentine Gallery. After Calmet.

the portraits of existing beings effigies which were at first intended to be merely emblematic and symbolical.

Early idolatry consisted, first, in separating the idea of the One Divinity into that of his various attributes, and of inventing symbols and making images of each separately; secondly, in the worship of the sun, moon, stars, and planets, as living existences; thirdly, in the deification of ancestors and early kings; and these three forms were often mingled together in strange and tangled confusion.

Amongst the famous personages with whose history men were made acquainted by oral tradition was Noah. He was known as the second father of the human race, and the preserver and teacher of the arts and sciences as they existed before the Great Deluge, of which so many separate traditions exist among the various races of mankind. Consequently, he was an object of worship in many countries and under many names; and his wife and sons, as his assistants in the diffusion of knowledge, were sometimes associated with him.

According to Berosus, of Babylon,—the Chaldean priest and astronomer, who extracted from the sacred books of "that great city" much interesting ancient lore, which he introduced into his 'History of Syria,' written, about B.C. 260, for the use of the Greeks,—at a time when men were sunk in barbarism, there came up from the Erythrean Sea (the Persian Gulf), and landed on the Babylonian shore, a creature named Oannes, which had the body and head of a fish. But above the fish's head was the head of a man, and below the tail of the fish were human feet. It had also human arms, a human voice, and human language. This strange monster sojourned among the rude people during the day, taking no food, but retiring to the sea at night; and it continued for some time thus to visit them, teaching them the arts of civilized life, and instructing them in science and religion.*

FIG. 2. — HEA, OR NOAH, THE GOD OF THE FLOOD. *Khorsabad.*

In this tale we have a distorted account of the life and occupation of Noah after his escape from the deluge which destroyed his home and drowned his neighbours. Oannes was one of the names under which

* Berosus, lib. i. p. 48.

he was worshipped in Chaldea, at Erech ("the place of the ark"), as the sacred and intelligent fish-god, the teacher of mankind, the god of science and knowledge. There he was also called Oes, Hoa, Ea, Ana, Anu, Aun, and Oan. Noah was worshipped, also, in Syria and Mesopotamia, and in Egypt, at "populous No,"* or Thebes—so named from "Theba," "the ark."

FIG. 3.—DAGON. *From a bas relief. Nimroud.*

The history of the coffin of Osiris is another version of Noah's ark, and the period during which that Egyptian divinity is said to have been shut up in it, after it was set afloat upon the waters, was precisely the same as that during which Noah remained in the ark.

Dagon, also — sometimes called Odacon—the great fish-god of the Philistines and Babylonians, was another phase of Oannes. "Dag," in Hebrew, signifies "a male fish," and "Aun" and "Oan" were two of the names of Noah. "Dag-aun" or "Dag-oan" therefore means "the fish Noah." He was portrayed in two ways. The more ancient image of him was that of a man issuing from a fish, as described of Oannes by Berosus; but in later times it was varied to that of a man whose upper half was human, and the lower parts those of

* Nahum iii. 8.

a fish. The image of Dagon which fell upon its face to the ground before "the ark of the God of Israel," was probably of this latter form, for we read * that in its fall, "the head of Dagon and both the palms of his hands were cut off upon the threshold: only the *stump* (in the margin, "*the fishy part*") of Dagon was left to him. This was evidently Milton's conception of him:

" Dagon his name; sea-monster,
    upward man
And downward fish." †

In some of the Nineveh sculptures of the fish-god, the head of the fish forms a kind of mitre on the head of the man, whilst the body of the fish appears as a cloak or cape over his shoulders and back. The fish varies in length; in some cases the tail almost touches the ground; in others it reaches but little below the man's waist.

FIG. 5.—DAGON. *From an Agate Signet. Nineveh.*

FIG. 4.—DAGON. *After Calmet.*

* 1 Samuel v. 4.
† 'Paradise Lost,' Book i. l. 462.

FIG. 6.—FISH AVATAR OF VISHNU.
After Calmet and Maurice.

In one of his "avatars," or incarnations, the god Vishnu "the Preserver," is represented as issuing from the mouth of a fish. He is celebrated as having miraculously preserved one righteous family, and, also, the Vedas, the sacred records, when the world was drowned. Not only is this legend of the Indian god wrought up with the history of Noah, but Vishnu and Noah bear the same name — Vishnu being the Sanscrit form of "Ish-nuh," "the man Noah." The word "avatar" also means "out of the boat." In fact the whole mythology of Greece and Rome, as well as of Asia, is full of the history and deeds of Noah, which it is impossible to misunderstand. In all the representations of a deity having a combined human and piscine form, the original idea was that of a person coming out of a fish—not being part of one, but issuing from it, as Noah issued from the ark. In all of them the fish denoted "preservation," "fecundity," "plenty," and "diffusion of knowledge."* As the image was not the effigy of a divine personage, but symbolized certain attributes of Divinity, its sex was comparatively unimportant, although it is possible that, combined with the fecundity of the fish, the idea of Noah's wife, as the second mother of all subsequent generations, according to the widely-spread and accepted traditions of the deluge, may have influenced the impersonation.

Atergatis, the far-famed goddess of the Syrians, was also a fish-divinity. Her image, like that of Dagon, had at first a fish's body with human extremities protruding from it; but in the course of centuries it was gradually altered to that of a being the upper portion of whose body was that of a woman and the lower half that of a fish. Gatis was a powerful queen of Sidon, and mother of Semiramis. She received the title of "Ater," or "Ader," "the Great," for the benefits she conferred on her people; one of these benefits being a strict conservation of their fisheries, both from their own imprudent use, and from foreign interference. She issued an edict that no fish should be eaten without her consent, and that no one should take fish in the neighbouring sea without a licence from herself. It is not improbable that she and her celebrated daughter, who

* Some writers are of the opinion that the legend of Oannes contains an allusion to the rising and setting of the sun, and that his semi-piscine form was the expression of the idea that half his time was spent above ground, and half below the waves. The same commentators also regard all the "civilizing" gods and goddesses as, respectively, solar and lunar deities. The attributes symbolized in the worship of Noah and the sun are so nearly alike that the two interpretations are not incompatible.

is said by Ovid and others to have been the builder of the walls of Babylon, were worshipped together; for that Atergatis was the same as the fish-goddess Ashteroth, or Ashtoreth, "the builder of the encompassing wall," we have, amongst other proofs, a remarkable one in Biblical history. In the first book of Maccabees v. 43, 44, we read that "all the heathen being discomfited before him (Judas Maccabeus) cast away their weapons, and fled unto the temple that was at *Carnaim*. But they took the city, and burned the temple with all that were therein. Thus was *Carnaim* subdued, neither could they stand any longer before Judas." In the second book of Maccabees xii. 26, we are told that "Maccabeus marched forth to *Carnion*, and to the temple of *Atargatis*, and there he slew five and twenty thousand persons." In Genesis xiv. 5, this city and temple are referred to as "*Ashteroth Karnaim*."

FIG. 7.—ATERGATIS.
From a Phœnician coin.

Fig. 7 is a representation of Atergatis on a medal coined at Marseilles. It shows that when the Phœnician colony from Syria, by whom that city was founded, settled there, they brought with them the worship of the gods of their country.

Atergatis was worshipped by the Greeks as Derceto and Astarte. Lucian writes*:—"In Phœnicia I saw the image of Derceto, a strange sight, truly! For she had the half of a woman, and from the thighs downwards a fish's tail." Diodorus Siculus describes (lib. ii.) the same deity, as represented at Ascalon, as "having the face of a woman,

---

* 'Opera Omnia,' tom. ii. p. 884, edit. Bened. de Deâ Syr.

but all the rest of the body a fish's." And this very same image at Ascalon, which Diodorus calls Derceto, or Atergatis, is denominated by Herodotus* "the celestial Aphrodite," who was identical with the Cyprian and Roman Venus. Of all the sacred buildings erected to the goddess, this temple was by far the most ancient; and the Cyprians

FIG. 8.—VENUS RISING FROM THE SEA, SUPPORTED BY TRITONS.
After Calmet.

themselves acknowledged that their temple was built after the model of it by certain Phœnicians who came from that part of Syria.

Thus the worship of Noah, as the second father of mankind, the repopulator of the earth, passed through various

* Lib. i. cap. cv.

phases and transformations till it merged in that of Venus, who rose from the sea, and was regarded as the representative of the reproductive power of Nature—the goddess whom Lucretius thus addressed:

> "Blest Venus! Thou the sea and fruitful earth
> Peoplest amain; to thee whatever lives
> Its being owes, and that it sees the sun:"

and to whom refers the passage in the Orphic hymn:

> "From thee are all things—all things thou producest
> Which are in heaven, or in the fertile earth,
> Or in the sea, or in the great abyss."

Under this latter phase—the impersonation of Venus—the fish portion of the body was discarded, and the cast-off form was allotted in popular credence to the Tritons—minor deities, who acknowledged the supremacy of the goddess, and were ready to render her homage and service by bearing her in their arms, drawing her chariot, etc., but who still possessed considerable power as sea-gods, and could calm the waves and rule the storm, at pleasure.

FIG. 9. FIG. 10.

VENUS DRAWN IN HER CHARIOT BY TRITONS. *From two Corinthian coins.*

Figs. 9 and 10 are from two Corinthian medals, each shewing Venus in a car or chariot drawn by Tritons, one male, the other female. On the obverse of Fig. 9, is the head of Nero, and on that of Fig. 10, the head of his grandmother Agrippina.*

From the very earliest period of history, then, the conjoined human and fish form was known to every generation of men. It was presented to their sight in childhood by sculptures and pictures, and was a conspicuous object in their religious worship. By the lapse of time its original import was lost and debased; and, from being

* It is worthy of note that the fish was also adopted as an emblem by the early Christians, and was frequently sculptured on their tombs as a private mark or sign of the faith in which the person there interred had died. It alluded to the letters which composed the Greek word Ιχθυς ("a fish") forming an anagram, the initials of words which conveyed the following sentiment: Ιησους, Jesus; Χριστος, Christ; Θεου, of God; Υιος, Son; Σωτηρ, Saviour. But it doubtless bore, also, the older meaning of "preservation" and "reproduction," of which the fish was the symbol, and betokened a belief in a future resurrection, as Noah was preserved to dwell in, and populate, a new world. In 'Sea Monsters Unmasked,' page 55, I gave a figure, copied by permission from the *Illustrated London News*, of a rough sculpture in the Roman catacombs, of Jonah being disgorged by a sea-monster. Near to it was found, on another Christian tomb, one of these designs of the "fish;" and it is not a little curious that, whereas the animal depicted as casting forth Jonah is not a whale, but a sea-serpent, or dragon, the *ichtheus* in this instance is apparently not a fish, but a seal.

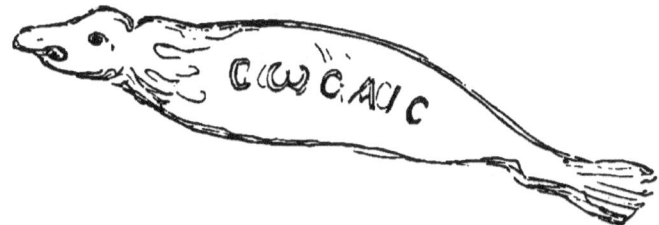

FIG. 11.—CHRISTIAN SYMBOL. *From the Catacombs at Rome.*

The article referred to appeared in the *Illustrated London News* of February 3rd, 1872, and the woodcut (fig. 11), an electrotype of which was most kindly presented to me by the proprietors of that paper, was one of the sketches that accompanied it.

an emblem and symbol, it came to be accepted as the corporeal shape and structure of actually-existent sea-deities, who might present themselves to the view of the mariner, in visible and tangible form, at any moment. Thus were men trained and prepared to believe in mermen and mermaids, to expect to meet with them at sea, and to recognise as one of them any animal the appearance and movements of which could possibly be brought into conformity with their pre-conceived ideas.

Accordingly, and very naturally, we find that from north to south this belief has been entertained. Megasthenes, who was a contemporary of Aristotle, but his junior, and whose geographical work was probably written at about the period of the great philosopher's death, reported that the sea which surrounded Taprobana, the ancient Ceylon, was inhabited by creatures having the appearance of women. Ælian stated that there were "whales," or "great fishes," having the form of satyrs. The early Portuguese settlers in India asserted that true mermen were found in the Eastern seas, and old Norse legends tell of submarine beings of conjoined human and piscine form, who dwell in a wide territory far below the region of the fishes, over which the sea, like the cloudy canopy of our sky, loftily rolls, and some of whom have, from time to time, landed on Scandinavian shores, exchanged their fishy extremities for human limbs, and acquired amphibious habits. Not only have poets sung of the wondrous and seductive beauty of the maidens of these aquatic tribes, but many a Jack tar has come home from sea prepared to affirm on oath that he has seen a mermaid. To the best of his belief he has told the truth. He has seen some living being which looked wonderfully human, and his imagination, aided by an inherited superstition, has supplied the rest.

THE MERMAID.

Before endeavouring to identify the object of his delusion, it may be well to mention a few instances of the supposed appearance of mermen and mermaidens in various localities.

Pliny writes *: "When Tiberius was emperor, an embassy was sent to him from Olysippo (Lisbon) expressly to inform him that a Triton, which was recognised as such by its form, had shown itself in a certain cave, and had been heard to produce loud sounds on a conch-shell. The Nereid, also, is not imaginary: its body is rough and covered with scales, but it has the appearance of a human being. For one was seen upon the same coast; and when it was dying those dwelling near at hand heard it moaning sadly for a long time. And the Governor of Gaul wrote to the divine Augustus that several Nereids had been found dead upon the shore. I have many informants—illustrious persons in high positions—who have assured me that they saw in the Sea of Cadiz a merman whose whole body was exactly like that of a man, that these mermen mount on board ships by night, and weigh down that end of the vessel on which they rest, and that if they are allowed to remain there long they will sink the ship."

Ælian in one of his short, jerky, disconnected chapters,† which rarely exceed a page in length, and some of which only contain two lines, writes: "It is reported that the great sea which surrounds the island of Taprobana (Ceylon) contains an immense multitude of fishes and whales, and some of them have the heads of lions, panthers, rams, and other animals; and (which is more wonderful still) some of the cetaceans have the form of satyrs. There are others which have the face of a woman, but prickles instead of hair. In addition to these, it is said there are other

---

* *Naturalis Historia*, Lib. ix. cap. v.
† *De Naturâ Animalium*, Lib. xvi. cap. xviii.

creatures of so strange and monstrous a kind that it would be impossible exactly to explain their appearance without the aid of a skilfully drawn picture: these have elongated and coiled tails, and, for feet, have claws * or fins. And I hear that in the same sea there are great amphibious beasts which are gregarious, and live on grain, and by night feed on the corn crops and grass, and are also very fond of the ripe fruit of the palms. To obtain these they encircle in their embrace the trees which are young and flexible, and, shaking them violently, enjoy the fruit which they thus cause to fall. When morning dawns they return to the sea, and plunge beneath the waves."

Ælian seems to have derived this information from Megasthenes, already referred to; but in another chapter,† he writes with greater certainty concerning these semi-human whales, and claims divine authority for his belief in the existence of tritons. "Although," he says, "we have no rational explanation nor absolute proof of that which fishermen are said to be able to affirm concerning the form of the tritons, we have the sworn testimony of many persons that there are in the sea cetaceans which from the head down to the middle of the body resemble the human species. Demostratus, in his works on fishing, says that an aged triton was seen near the town of Tanagra, in Bœotia, which was like the drawings and pictures of tritons, but its features were so obscured by age, and it disappeared so quickly, that its true character was not easily perceptible. But on the spot where it had rested on the shore were found some rough and very hard scales which had become detached from it. A certain senator—one of those selected by lot to carry on

---

* "*Forfices*," literally "shears," or "nippers," like the claws of a lobster.

† Lib. xiii. cap. xxi.

the administration of Achaia and the duties of the annual magistracy" (the mayor, in fact,) "being anxious to investigate the nature of this triton, put a portion of its skin on the fire. It gave out a most horrible odour; and those standing by were unable to decide whether it belonged to a terrestrial or marine animal. But the magistrate's curiosity had an evil ending, for very soon afterwards, whilst crossing a narrow creek in a boat, he fell overboard and was drowned; and the Tanagreans all regarded this as a judgment upon him for his crime of impiety towards the triton—an interpretation which was confirmed when his decomposing body was cast ashore, for it emitted exactly the same odour as had the burned skin of the triton. The Tanagreans and Demostratus explain whence the triton had strayed, and how it was stranded in this place. I believe," continues Ælian, "that tritons exist, and I reverentially produce as my witness a most veracious god—namely, Apollo Didymæus, whom no man in his senses would presume to regard as unworthy of credit. He sings thus of the triton, which he calls the sheep of the sea:

> '*Dum vocale maris monstrum natat æquore triton,*
> *Neptuni pecus, in funes forte incidit extra*
> *Demissos navim';*"

which I venture to translate as follows:

> A triton, vocal monster of the deep,
> One of a flock of Neptune's scaly sheep,
> Was caught, whilst swimming o'er the watery plain,
> By lines which fishers from their boat had lain.

"Therefore," Ælian concludes, "if he, the omniscient god, pronounces that there are tritons, it does not behove us to doubt their existence."

Sir J. Emerson Tennent, in his 'Natural History of Ceylon,' quoting from the *Histoire de la Compagnie de*

Jesus, mentions that the annalist of the exploits of the Jesuits in India gravely records that seven of these monsters, male and female, were captured at Manaar, in 1560, and carried to Goa, where they were dissected by Demas Bosquez, physician to the Viceroy, "and their internal structure found to be in all respects conformable to the human." He also quotes Valentyn, one of the Dutch colonial chaplains, who, in his account of the Natural History of Amboyna,* embodied in his great work on the Netherlands' possessions in India, published in 1727,† devoted the first section of his chapter on the fishes of that island to a minute description of the "Zee-Menschen," "Zee-Wyven," and mermaids, the existence of which he warmly insists on as being beyond cavil. He relates that in 1663, when a lieutenant in the Dutch service was leading a party of soldiers along the sea-shore in Amboyna, he and all his company saw the mermen swimming at a short distance from the beach. They had long and flowing hair of a colour between grey and green. Six weeks afterwards the creatures were again seen by him and more than fifty witnesses, at the same place, by clear. daylight. "If any narrative in the world," adds Valentyn, "deserves credit it is this; since not only one, but two mermen together were seen by so many eye-witnesses. Should the stubborn world, however, hesitate to believe it, it matters nothing, as there are people who would even deny that such cities as Rome, Constantinople, or Cairo, exist, merely because they themselves have not happened to see them. But what are such incredulous persons," he continues, "to make

---

* One of the Dutch spice-islands in the Banda Sea, between Celebes and Papua.

† *Beschrijving van Oud en Nieuw Oost-Indien*, etc., 5 vols. folio, Dordrecht and Amsterdam, 1727, vol. iii. p. 330.

of the circumstance recorded by Albrecht Herport * in his account of India, that a merman was seen in the water near the church of Taquan on the morning of the 29th of April, 1661, and a mermaid at the same spot the same afternoon? Or what do they say to the fact that in 1714 a mermaid was not only seen but captured near the island of Booro, five feet, Rhineland measure, in height; which lived four days and seven hours, but, refusing all food, died without leaving any intelligible account of herself?"

FIG. 12.—MERMAID AND FISHES OF AMBOYNA. *After Valentyn.*

Valentyn, in support of his own faith in the mermaid, cites many other instances in which both "sea-men and sea-women" were seen and taken at Amboyna; especially one by a district visitor of the church, who presented it to the Governor Vanderstel. Of this "well-authenticated" specimen he gives an elaborate portrait amongst the fishes of the island,†

* *Itinerarium Indicum*, Berne, 1669.
† With the permission and assistance of Messrs. Longman, the accompanying wood-cut of this picture, and that of the Dugong, on page 43, are copied from Sir J. Emerson Tennent's book published in 1861.

with a minute description of each for the satisfaction of men of science.

The fame of this creature having reached Europe, the British minister in Holland wrote to Valentyn on the 28th of December, 1716, whilst the Emperor Peter the Great, of Russia, was his guest at Amsterdam, to communicate the desire of the Czar that the mermaid should be brought home from Amboyna for his inspection. To complete his proofs of the existence of mermen and merwomen, Valentyn points triumphantly to the historical fact that in Holland, in the year 1404, a mermaid was driven, during a tempest, through a breach in the dyke of Edam, and was taken alive in the lake of Purmer. Thence she was carried to Haarlem, where the Dutch women taught her to spin, and where several years after, she died in the Roman Catholic faith;— "but this," says the pious Calvinistic chaplain, "in no way militates against the truth of her story." The worthy minister citing the authority of various writers as proof that mermaids had in all ages been known in Gaul, Naples, Epirus, and the Morea, comes to the conclusion that as there are "sea-cows," "sea-horses," "sea-dogs," as well as "sea-trees," and "sea-flowers," which he himself had seen, there are no reasonable grounds for doubt that there may also be "sea-maidens" and "sea-men."

In an early account of Newfoundland,* Whitbourne describes a "maremaid or mareman," which he had seen "within the length of a pike," and which "came swimming swiftly towards him, looking cheerfully on his face, as it had been a woman. By the face, eyes, nose, mouth, chin, ears, neck and forehead, it appeared to be so beautiful, and in those parts so well proportioned, having round about the head many blue streaks resembling hair, but certainly it

* Whitbourne's 'Discourse of Newfoundland.'

was no hair. The shoulders and back down to the middle were square, white, and smooth as the back of a man, and from the middle to the end it tapered like a broad-hooked arrow." The animal put both its paws on the side of the boat wherein its observer sat, and strove much to get in, but was repelled by a blow.

In 1676, a description was given by an English surgeon named Glover, of an animal of this kind. The author did not designate it by any name, but the incident has the honour of being recorded in the *Philosophical Transactions.** About three leagues from the mouth of the river Rappahannock, in America, while alone in a vessel, he observed, at the distance of about half a stone-throw, he says, "a most prodigious creature, much resembling a man, only somewhat larger, standing right up in the water, with his head, neck, shoulders, breast and waist, to the cubits of his arms, above water, and his skin was tawny, much like that of an Indian; the figure of his head was pyramidal and sleek, without hair; his eyes large and black, and so were his eyebrows; his mouth very wide, with a broad black streak on the upper lip, which turned upwards at each end like mustachios. His countenance was grim and terrible. His neck, shoulders, arms, breast and waist, were like unto the neck, arms, shoulders, breast and waist of a man. His hands, if he had any, were under water. He seemed to stand with his eyes fixed on me for some time, and afterwards dived down, and, a little after, rose at somewhat a greater distance, and turned his head towards me again, and then immediately fell a little under water, that I could discern him throw out his arms and gather them in as a man does when he swims. At last, he shot with his head downwards, by which means he cast his tail above the

* Glover's 'Account of Virginia,' ap. Phil. Trans. vol. xi. p. 625.

water, which exactly resembled the tail of a fish, with a broad fane at the end of it."

Thormodus Torfæus [*] maintains that mermaids are found on the south coast of Iceland, and, according to Olafsen,[†] two have been taken in the surrounding seas, the first in the earlier part of the history of that island, and the second in 1733. The latter was found in the stomach of a shark. Its lower parts were consumed, but the upper were entire. They were as large as those of a boy eight or nine years old. Both the cutting teeth and grinders were long and shaped like pins, and the fingers were connected by a large web. Olafsen was inclined to believe that these were human remains, but the islanders all firmly maintained that they were part of "a marmennill," by which name the mermaid is known among them.

Of course the worthy bishop of Bergen, Pontoppidan, has something to tell us about mermaids in his part of the world. "Amongst the sea monsters," he says, [‡] "which are in the North Sea, and are often seen, I shall give the first place to the Hav-manden, or merman, whose mate is called Hav-fruen, or mermaid. The existence of this creature is questioned by many, nor is it at all to be wondered at, because most of the accounts we have had of it are mixed with mere fables, and may be looked upon as idle tales." As such he regards the story told by Jonas Ramus in his 'History of Norway,' of a mermaid taken by fishermen at Hordeland, near Bergen, and which is said to have sung an unmusical song to King Hiorlief. In the same category he places an account given by Besenius in his life of Frederic II. (1577), of a mermaid that called

[*] *Historia rerum Norvegicarum.*
[†] *Voyage en Islande,* tom. iii. p. 223.
[‡] 'Natural History of Norway,' vol. ii. p. 190.

herself Isbrandt, and held several conversations with a peasant at Samsoe, in which she foretold the birth of King Christian IV., "and made the peasant preach repentance to the courtiers, who were very much given to drunkenness." Equally "idle" with the above stories is, in his opinion, another, extracted from an old manuscript still to be seen in the University Library at Copenhagen, and quoted by Andrew Bussæus (1619), of a merman caught by the two senators, Ulf Rosensparre and Christian Holch, whilst on their voyage home to Denmark from Norway. This sea-man frightened the two worshipful gentlemen so terribly that they were glad to let him go again; for as he lay upon the deck he spoke Danish to them, and threatened that if they did not give him his liberty "the ship should be cast away, and every soul of the crew should perish."

"When such fictions as these," says Pontoppidan, "are mixed with the history of the merman, and when that creature is represented as a prophet and an orator; when they give the mermaid a melodious voice, and tell us that she is a fine singer, we need not wonder that so few people of sense will give credit to such absurdities, or that they even doubt the existence of such a creature." The good prelate, however, goes on to say that "whilst we have no ground to believe all these fables, yet, as to the existence of the creature we may safely give our assent to it," and, "if this be called in question, it must proceed entirely from the fabulous stories usually mixed with the truth." Like Valentyn, he argues that as there are "sea-horses," "sea-cows," "sea-wolves," "sea-dogs," "sea-hogs," etc., it is probable from analogy, that "we should find in the ocean a fish or creature which resembles the human species more than any other." As for the objection "founded on self-love and respect to our own species which is honoured with the image of God, who made man lord of all creatures, and that, consequently, we may suppose he is entitled to a noble and heavenly form which other creatures must not partake of," he thinks "its force vanishes when we consider the form of apes, and especially of another African creature called 'Quoyas Morrov' described by Odoard Dapper" in his work on Africa, and which appears to have been a chimpanzee. Pontoppidan regarded it as being the Satyr of the ancients. He therefore claims that "if we will not allow our Norwegian Hastromber the honourable name of merman, we may very well call it the 'Sea-ape,' or the 'Sea-Quoyas-Morrov;' especially as the author already quoted says that, "in the Sea of Angola mermaids are frequently caught which resemble the human species. They are taken in nets, and killed by the negroes, and are heard to shriek and cry like women."

The Bishop adds that in the diocese of Bergen, as well as in the manor of Nordland, there were hundreds of persons who affirmed with the strongest assurances that they had seen this kind of creature; sometimes at a distance and at other times quite close to their boats, standing upright, and formed like a human creature down to the middle—the rest they could not see—but of those who had seen them out of water and handled them he had not been able to find more than one person of credit who could vouch it for truth. This informant, "the Reverend Mr. Peter Angel, minister of Vand-Elvens Gield, on Suderoe," assured his bishop, when he was on a visitation journey, that "in the year 1719, he (being then about twenty years old) saw what is called a merman lying dead on a point of land near the sea, which had been cast ashore by the waves along with several sea-calves (seals), and other dead fish.

The length of this creature was much greater than what has been mentioned of any before, namely, above three fathoms. It was of a dark grey colour all over: in the lower part it was like a fish, and had a tail like that of a porpoise. The face resembled that of a man, with a mouth, forehead, eyes, etc. The nose was flat, and, as it were, pressed down to the face, in which the nostrils were very visible. The breast was not far from the head; the arms seemed to hang to the side, to which they were joined by a thin skin, or membrane. The hands were, to all appearance, like the paws of a sea-calf. The back of this creature was very fat, and a great part of it was cut off, which, with the liver, yielded a large quantity of train-oil." The author then quotes a description by Luke Debes [*] of a mermaid seen in 1670 at Faroe, westward of Qualboe Eide, by many of the inhabitants, as also by others from different parts of Suderoe. She was close to the shore, and stood there for two hours and a half, and was up to her waist in water. She had long hairs on her head, which hung down to the surface of the water all round about her, and she held a fish in her right hand.

Pontoppidan mentions other instances of similar appearances, and says that the latest he had heard of was of a merman seen in Denmark on the 20th of September, 1723, by three ferrymen who, at some distance from the land, were towing a ship just arrived from the Baltic. Having caught sight of something which looked like a dead body floating on the water, they rowed towards it, and there, resting on their oars, allowed it to drift close to them. It sank, but immediately came to the surface again, and then they saw that it had the appearance of an old man, strong-

[*] *Feroa Reserata*, or Description of the Feroe Islands. 8vo. Copenhagen, 1673.

limbed, and with broad shoulders, but his arms they could not see. His head was small in proportion to his body, and had short, curled, black hair, which did not reach below his ears; his eyes lay deep in his head, and he had a meagre and pinched face, with a black, coarse beard, that looked as if it had been cut. His skin was coarse, and very full of hair. He stood in the same place for half a quarter of an hour, and was seen above the water down to his breast: at last the men grew apprehensive of some danger, and began to retire; upon which the monster blew up his cheeks, and made a kind of roaring noise, and then dived under water, so that they did not see him any more. One of them, Peter Gunnersen, related (what the others did not observe) that this merman was, about the body and downwards, quite pointed, like a fish. This same Peter Gunnersen likewise deposed that "about twenty years before, as he was in a boat near Kulleor, the place where he was born, he saw a mermaid with long hair and large breasts." He and his two companions were, by command of the king, examined by the burgomaster of Elsineur, Andrew Bussæus, before the privy-councillor, Fridrich von Gram, and their testimony to the above effect was given on their respective oaths.

Brave old Henry Hudson, the sturdy and renowned navigator, who thrice, in three successive years, gave battle to the northern ice, and was each time defeated in his endeavour to discover a north-west or north-east passage to China, though he stamped his name on the title-page of a mighty nation's history, records the following incident: "This evening (June 15th) one of our company, looking overboard, saw a mermaid, and, calling up some of the company to see her, one more of the crew came up, and by that time she was come close to the ship's side, looking earnestly on the men. A little after a sea came and overturned her. From the navel upward, her back and breasts were like a woman's, as they say that saw her; her body as big as one of us, her skin very white, and long hair hanging down behind, of colour black. In her going down they saw her tail, which was like the tail of a porpoise and speckled like a mackarel's. Their names that saw her were Thomas Hilles and Robert Rayner."

Steller, who was a zoologist of some repute, reports having seen in Behrings Straits a strange animal, which he calls a "sea-ape," and in which one might almost recognise Pontoppidan's "Sea-Quoyas-Morrov." It was about five feet long, had sharp and erect ears and large eyes, and on its lips a kind of beard. Its body was thick and round, and it tapered to the tail, which was bifurcated, with the upper lobe longest. It was covered with thick hair, grey on the back, and red on the belly. No feet nor paws were visible. It was full of frolic, and sported in the manner of a monkey, swimming sometimes on one side of the ship and sometimes on the other. It often raised one-third of its body out of the water, and stood upright for a considerable time. It would frequently bring up a sea-plant, not unlike a bottle-gourd, which it would toss about and catch in its mouth, playing numberless fantastic tricks with it.

Somewhat similar accounts have been brought from the Southern Hemisphere, two, at least, of which are worth transcribing.

Captain Colnett, in his 'Voyage to the South Atlantic,' says:—"A very singular circumstance happened off the coast of Chili, in lat. 24° S., which spread some alarm amongst my people, and awakened their superstitious apprehensions. About 8 o'clock in the evening an animal rose alongside the ship, and uttered such shrieks and tones of lamentation, so much like those produced by the female human voice when expressing the deepest distress as to occasion no small degree of alarm among those who first heard it. These cries continued for upwards of three hours, and seemed to increase as the ship sailed from it. I never heard any noise whatever that approached so near those sounds which proceed from the organs of utterance in the human species."

Captain Weddell, in his 'Voyage towards the South Pole' (p. 143), writes that one of his men, having been left ashore on Hall's Island to take care of some produce, heard one night about ten o'clock, after he had lain down to rest, a noise resembling human cries. As daylight does not disappear in those latitudes at the season in which the incident occurred, the sailor rose and searched along the beach, thinking that, possibly, a boat might have been upset, and that some of the crew might be clinging to the detached rocks.

> "Roused by that voice of silver sound,
> From the paved floor he lightly sprung,
> And, glaring with his eyes around,
> Where the fair nymph her tresses wrung," *

guided by occasional sounds, he at length saw an object lying on a rock a dozen yards from the shore, at which he was somewhat frightened. "The face and shoulders appeared of human form and of a reddish colour; over the shoulders hung long green hair; the tail resembled that of a seal, but the extremities of the arms he could not see distinctly."

> "As on the wond'ring youth she smiled,
> Again she raised the melting lay,"*

---

* John Leyden.

for the creature continued to make a musical noise during the two minutes he gazed at it, and, on perceiving him, disappeared in an instant.

The universality of the belief in an animal of combined human and fish-like form is very remarkable. That it exists amongst the Japanese we have evidence in their curious and ingeniously-constructed models which are occasionally brought to this country. I have one of these which is so exactly the counterpart of that which my friend Mr. Frank Buckland described, originally in *Land and Water*, and which forms the subject of a chapter in his 'Curiosities of Natural History,'* that the portrait of the one (Fig. 13) will equally well represent

FIG. 13.—A JAPANESE ARTIFICIAL MERMAID.

the other. The lower half of the body is made of the skin and scales of a fish of the carp family, and fastened on to this, so neatly that it is hardly possible to detect where the joint is made, is a wooden body, the ribs of which are so prominent that the poor mermaid has a miserable and half-starved appearance. The upper part of the body is in the attitude of a Sphinx, leaning upon its elbows and fore-arms. The arms are thin and scraggy, and the fingers attenuated and skeleton-like. The nails are formed of small pieces of

* Third Series, vol. ii. p. 134, 2nd ed.

ivory or bone. The head is like that of a small monkey, and a little wool covers the crown, so thinly and untidily that if the mermaid possessed a crystal mirror she would see the necessity for the vigorous use of her comb of pearl. The teeth are those of some fish—apparently of the cat-fish, (*Anarchicas lupus*). These Japanese artificial mermaids have brought many a dollar into the pockets of Mr. Barnum and other showmen.

Somewhat different in appearance from this, but of the same kind, was an artificial mermaid described in the *Saturday Magazine* of June 4th, 1836. Fig. 14 is a facsimile of the woodcut which accompanied it. This grotesque composition was exhibited in a glass case, some years previously, "in a leading street at the west end" of London. It was constructed "of the skin of the head and shoulders of a monkey, which was attached to the dried skin of a fish of the salmon kind with the head cut off, and the whole was stuffed and highly varnished, the better to deceive the eye." It was said to have been "taken by the crew of a Dutch vessel from on board a native Malacca boat, and from the reverence shown to it, it was supposed to be a representative of one of their idol gods." I am inclined to think that it was of Japanese origin.

FIG. 14.—AN ARTIFICIAL MERMAID, PROBABLY JAPANESE.

Fig. 15 is described in the article above referred to as having been copied from a Japanese drawing, and as being a portrait of one of their deities. Its similarity to one of those of the Assyrians (Fig. 2, page 3) is remarkable. The inscription, however, does not indicate this. The Chinese characters in the centre—"*Nin giyo*"—signify "human fish;" those on the right in Japanese *Hira Kana*, or running-hand, have the same purport, and those on the left, in *Kata Kana*, the characters of the Japanese alphabet, mean "*Ichi hiru ike*"—"one day kept alive." The whole legend seems to pretend that this human fish was actually caught, and kept alive in water for twenty-four hours, but, as the box on which it is inscribed is one of those in which the Japanese showmen keep their toys, it was probably the subject of a "penny peep-show."

We need not travel from our own country to find the belief in mermaids yet existing. It is still credited in the north of Scotland that they inhabit the neighbouring seas: and Dr. Robert Hamilton, F.R.S.E., writing in 1839, expressed emphatically his opinion that there was then as much ignorance on this subject as had prevailed at any former period.*

FIG. 15.—A MERMAID. *From a Japanese picture.*

In the year 1797, Mr. Munro, schoolmaster of Thurso, affirmed that he had seen "a figure like a naked female, sitting on a rock projecting into the sea, at Sandside Head, in the parish of Reay. Its head was covered with long, thick, light-brown hair, flowing down on the shoulders. The forehead was round, the face plump, and the cheeks ruddy. The mouth and lips resembled those of a human being, and the eyes were blue. The arms, fingers, breast,

* Naturalist's Library, Marine Amphibiæ, p. 291.

and abdomen were as large as those of a full-grown female," and, altogether,

> "That sea-nymph's form of pearly light
> Was whiter than the downy spray,
> And round her bosom, heaving bright,
> Her glossy yellow ringlets play." *

"This creature," continued Mr. Munro, "was apparently in the act of combing its hair with its fingers, which seemed to afford it pleasure, and it remained thus occupied during some minutes, when it dropped into the sea." The Dominie

> "saw the maiden there,
> Just as the daylight faded,
> Braiding her locks of gowden hair
> An' singing as she braided," †

but he did not remark whether the fingers were webbed. On the whole, he infers that this was a marine animal of which he had a distinct and satisfactory view, and that the portion seen by him bore a narrow resemblance to the human form. But for the dangerous situation it had chosen, and its appearance among the waves, he would have supposed it to be a woman. Twelve years later, several persons observed near the same spot an animal which they also supposed to be a mermaid.

A very remarkable story of this kind is one related by Dr. Robert Hamilton in the volume already referred to, and for the general truth of which he vouches, from his personal knowledge of some of the persons connected with the occurrence. In 1823 it was reported that some fishermen of Yell, one of the Shetland group, had captured a mermaid by its being entangled in their lines. The statement was that

* John Leyden.
† The Ettrick Shepherd.

"the animal was about three feet long, the upper part of the body resembling the human, with protuberant mammæ, like a woman; the face, forehead, and neck were short, and resembled those of a monkey; the arms, which were small, were kept folded across the breast; the fingers were distinct, not webbed; a few stiff, long bristles were on the top of the head, extending down to the shoulders, and these it could erect and depress at pleasure, something like a crest. The inferior part of the body was like a fish. The skin was smooth, and of a grey colour. It offered no resistance, nor attempted to bite, but uttered a low, plaintive sound. The crew, six in number, took it within their boat, but, superstition getting the better of curiosity, they carefully disentangled it from the lines and a hook which had accidentally become fastened in its body, and returned it to its native element. It instantly dived, descending in a perpendicular direction." Mr. Edmonston, the original narrator of this incident, was "a well-known and intelligent observer," says Dr. Hamilton, and in a communication made by him to the Professor of Natural History in the Edinburgh University gave the following additional particulars, which he had learned from the skipper and one of the crew of the boat. "They had the animal for three hours within the boat: the body was without scales or hair; it was of a silvery grey colour above, and white below; it was like the human skin; no gills were observed, nor fins on the back or belly. The tail was like that of a dog-fish; the mammæ were about as large as those of a woman; the mouth and lips were very distinct, and resembled the human. Not one of the six men dreamed of a doubt of its being a mermaid, and it could not be suggested that they were influenced by their fears, for the mermaid is not an object of terror to fishermen: it is rather a welcome guest, and danger is apprehended from its experiencing bad treatment." Mr. Edmonston concludes by saying that "the usual resources of scepticism that the seals and other sea-animals appearing under certain circumstances, operating upon an excited imagination, and so producing ocular illusion, cannot avail here. It is quite impossible that six Shetland fishermen could commit such a mistake." It would seem that the narrator demands that his readers shall be silenced, if unconvinced; but

> "He that complies against his will
> Is of his own opinion still."

This incident is well-attested, and merits respectful and careful consideration; but I decline to admit any such impossibility of error in observation or description on the part of the fishermen, or the further impossibility of recognising in the animal captured by them one known to naturalists. The particulars given in this instance, and also of the supposed merman seen cast ashore dead in 1719 by the Rev. Peter Angel (p. 22), are sufficiently accurate descriptions of a warm-blooded marine animal, with which the Shetlanders, and probably Mr. Edmonston also, were unacquainted, namely, the rytina, of which I shall have more to say presently; and these occurrences afford some slight hope that this remarkable beast may not have become extinct in 1768, as has been supposed, but that it may still exist somewhat further south than it was met with by its original describer, Steller.

Turning to Ireland, we find the same credence in the semi-human fish, or fish-tailed human being. In the autumn of 1819 it was affirmed that "a creature appeared on the Irish coast, about the size of a girl ten years of age, with a bosom as prominent as one of sixteen, having a profusion of long dark-brown hair, and full, dark eyes. The hands and arms were formed like those of a man, with a slight web connecting the upper part of the fingers, which were frequently employed in throwing back and dividing the hair. The tail appeared like that of a dolphin." This creature remained basking on the rocks during an hour, in the sight of numbers of people, until frightened by the flash of a musket, when

> "Away she went with a sea-gull's scream,
> And a splash of her saucy tail,"*

for it instantly plunged with a scream into the sea.

From Irish legends we learn that those sea-nereids, the "Merrows," or "Moruachs" came occasionally from the sea, gained the affections of men, and interested themselves in their affairs; and similar traditions of the "Morgan" (sea-women) and the "Morverch" (sea-daughters) are current in Brittany.

In English poetry the mermaid has been the subject of many charming verses, and Shakspeare alludes to it in his plays no less than six times. The head-quarters of these "daughters of the sea" in England, or of the belief in their existence, are in Cornwall. There the fisherman, many a time and

> "Oft, beneath the silver moon,†
> Has heard, afar, the mermaid sing,"

and has listened, so they say, to

> "The mermaid's sweet sea-soothing lay
> That charmed the dancing waves to sleep."†

---

* Tom Hood. 'The Mermaid at Margate.'
† John Leyden.

Mr. Robert Hunt, F.R.S., in his collection of the traditions and superstitions of old Cornwall,* records several curious legends of the "merrymaids" and "merrymen" (the local name of mermaids), which he had gathered from the fisher-folk and peasants in different parts of that county.

And, in a pleasant article in 'All the Year Round,'† 1865, "A Cornish Vicar"‡ mentions some of the superstitions of the people in his neighbourhood, and the perplexing questions they occasionally put to him. One of his parishioners, an old man named Anthony Cleverdon, but who was popularly known as "Uncle Tony," having been the seventh son of his parents, in direct succession, was looked upon, in consequence, as a soothsayer. This "ancient augur" confided to his pastor many highly efficacious charms and formularies, and, in return, sought for information from him on other subjects. One day he puzzled the parson by a question which so well illustrates the local ideas concerning mermaids, and the sequel of which is, moreover, so humorously related by the vicar, that I venture to quote his own words, as follows:—

"Uncle Tony said to me, 'Sir, there is one thing I want to ask you, if I may be so free, and it is this: why should a merrymaid, that will ride about upon the waters in such terrible storms, and toss from sea to sea in such ruckles as there be upon the coast, why should she never lose her looking-glass and comb?' 'Well, I suppose,' said I, 'that if there are such creatures, Tony, they must wear their looking-glasses and combs fastened on somehow, like fins

* 'Romances and Drolls of the West of England.' London: Hotten, 1871.
† Vol. xiii. p. 336.
‡ The "Cornish Vicar" was, evidently, the Rev. Robert Stephen Hawker, M.A., Vicar of Morwenstow, and author of 'Echoes from Old Cornwall,' 'Footprints of Former Men in Cornwall,' etc.

to a fish.' 'See!' said Tony, chuckling with delight, 'what a thing it is to know the Scriptures, like your reverence; I should never have found it out. But there's another point, sir, I should like to know, if you please; I've been bothered about it in my mind hundreds of times. Here be I, that have gone up and down Holacombe cliffs and streams fifty years come next Candlemas, and I've gone and watched the water by moonlight and sunlight, days and nights, on purpose, in rough weather and smooth (even Sundays, too, saving your presence), and my sight as good as most men's, and yet I never could come to see a merrymaid in all my life: how's that, sir?' 'Are you sure, Tony,' I rejoined, 'that there are such things in existence at all?' 'Oh, sir, my old father seen her twice! He was out one night for wreck (my father watched the coast, like most of the old people formerly), and it came to pass that he was down at the duck-pool on the sand at low-water tide, and all to once he heard music in the sea. Well, he croped on behind a rock, like a coastguardsman watching a boat, and got very near the music .... and there was the merrymaid, very plain to be seen, swimming about upon the waves like a woman bathing—and singing away. But my father said it was very sad and solemn to hear—more like the tune of a funeral hymn than a Christmas carol, by far—but it was so sweet that it was as much as he could do to hold back from plunging into the tide after her. And he an old man of sixty-seven, with a wife and a houseful of children at home. The second time was down here by Holacombe Pits. He had been looking out for spars— there was a ship breaking up in the Channel—and he saw some one move just at half-tide mark, so he went on very softly, step by step, till he got nigh the place, and there was the merrymaid sitting on a rock, the bootyfullest merrymaid that eye could behold, and she was twisting about her long hair, and dressing it, just like one of our girls getting ready for her sweetheart on the Sabbath-day. The old man made sure he should greep hold of her before ever she found him out, and he had got so near that a couple of paces more and he would have caught her by the hair, as sure as tithe or tax, when, lo and behold, she looked back and glimpsed him! So, in one moment she dived head-foremost off the rock, and then tumbled herself topsy-turvy about in the water, and cast a look at my poor father, and grinned like a seal.'" And a seal it probably was that Tony's "poor father" saw.

What, then, are these mermaids and mermen, a belief in whose existence has prevailed in all ages, and amongst all the nations of the earth? Have they, really, some of the parts and proportions of man, or do they belong to another order of mammals on which credulity and inaccurate observation have bestowed a false character?

Mr. Swainson, a naturalist of deserved eminence, has maintained on purely scientific grounds, that there must exist a marine animal uniting the general form of a fish with that of a man; that by the laws of Nature the natatorial type of the *Quadrumana* is most assuredly wanting, and that, apart from man, a being connecting the seals with the monkeys is required to complete the circle of quadrumanous animals.*

Mr. Gosse † argues that all the characters which Mr. Swainson selects as marking the natatorial type of animals belong to man, and that he being, in his savage state, a great swimmer, is the true aquatic primate, which Mr. Swainson regards as absent. Mr. Gosse admits, however, that "nature

---

* 'Geography and Distribution of Animals.'
† 'Romance of Natural History,' 2nd Series.

has an odd way of mocking at our impossibilities, and " that "it *may be* that green-haired maidens with oary tails, lurk in the ocean caves, and keep mirrors and combs upon their rocky shelves ;" and the conclusion he arrives at is that the combined evidence "induces a strong suspicion that the northern seas may hold forms of life as yet uncatalogued by science."

That there are animals in the northern and other seas with which we are unacquainted, is more than probable : discoveries of animals of new species are constantly being made, especially in the life of the deep sea. But I venture to think that the production of an animal at present unknown is quite unnecessary to account for the supposed appearances of mermaids.

We have in the form and habits of the *Phocidæ*, or earless seals, a sufficient interpretation of almost every incident of the kind that has occurred north of the Equator—of those in which protuberant *mammæ* are described, we must presently seek another explanation. The round, plump, expressive face of a seal, the beautiful, limpid eyes, the hand-like fore-paws, the sleek body, tapering towards the flattened hinder fins, which are directed backwards, and spread out in the form of a broad fin, like the tail of a fish, might well give the idea of an animal having the anterior part of its body human and the posterior half piscine.

In the habits of the seals, also, we may trace those of the supposed mermaid, and the more easily the better we are acquainted with them. All seals are fond of leaving the water frequently. They always select the flattest and most shelving rocks which have been covered at high tide, and prefer those that are separated from the mainland. They generally go ashore at half-tide, and invariably lie with their heads towards the water, and seldom more than a yard or two from it. There they will often remain, if undisturbed, for six hours; that is, until the returning tide floats them off the rock. As for the sweet melody, "so melting soft," that must depend much on the ear and musical taste of the listener. I have never heard a seal utter any vocal sounds but a porcine grunt, a plaintive moan, and a pitiful whine. But another habit of the seals has, probably more than anything else, caused them to be mistaken for semi-human beings,—namely, that of poising themselves upright in the water with the head and the upper third part of the body above the surface.

One calm sunny morning in August, 1881, a fine schooner-yacht, on board of which I was a guest, was slowly gliding out of the mouth of the river Maas, past the Hook of Holland, into the North Sea, when a seal rose just ahead of us, and assumed the attitude above described. It waited whilst we passed it, inspecting us apparently with the greatest interest; then dived, swam in the direction in which we were sailing, so as to intercept our course, and came up again, sitting upright as before. This it repeated three times, and so easily might it have been taken for a mermaid, that one of the party, who was called on deck to see it, thought, at first, that it was a boy who had swam off from the shore to the vessel on a begging expedition.

Laing, in his account of a voyage to the North, mentions having seen a seal under similar circumstances.

A young seal which was brought from Yarmouth to the Brighton Aquarium in 1873, habitually sat thus, showing his head and a considerable portion of his body out of water. His bath was so shallow in some parts that he was able to touch the bottom, and, with his after-flippers tucked under him, like a lobster's tail, and spread out in front, he would balance himself on his hind quarters, and look inquisitively at everybody, and listen attentively to everything within sight and hearing. When he was satisfied that no one was likely to interfere with him, and that it was unnecessary to be on the alert, he would half-close his beautiful, soft eyes, and either contentedly pat, stroke, and scratch his little fat stomach with his right paw, or flap both of them across his breast in a most ludicrous manner, exactly as a cabman warms the tips of his fingers on a wintry day, by swinging his arms vigorously across his chest, and striking his hands against his body on either side. He was very sensitive to musical sounds, as many dogs are, and when a concert took place in the building a high note from one of the vocalists would cause him to utter a mournful wail, and to dive with a splash that made the water fly, the audience smile, and the singer frown.

Captain Scoresby tells us that he had seen the walrus with its head above water, and in such a position that it required little stretch of imagination to mistake it for a human being, and that on one occasion of this kind the surgeon of his ship actually reported to him that he had seen a man with his head above water.

Peter Gunnersen's merman (p. 24), who "blew up his cheeks and made a kind of roaring noise" before diving, was probably a "bladder-nose" seal. The males of that species have on the head a peculiar pad, which they can dilate at pleasure, and their voice is loud and discordant.

The appearance and behaviour of Steller's "sea-ape," described on p. 25, may, I think, be attributed to one of the eared seals, the so-called sea-lions, or sea-bears. Every one who has seen these animals fed must have noticed the rapidity with which they will dive and swim to any part of their pond where they expect to receive food, and how, like a dog after a pebble, they will keenly watch their keeper's movements, and start in the direction to which he is apparently about to throw a fish, even before the latter has left his hand. This may be seen at the Zoological Gardens, Regent's Park, and, better than anywhere else in Europe, at the Jardin d'Acclimatation, Paris. It would be quite in accordance with their habits that one of these *Otaria* should dive under a ship, and rise above the surface on either side, eagerly surveying those on board, in hope of obtaining food, or from mere curiosity.

The seals and their movements account for so many mermaid stories, that all accounts of sea-women with prominent bosoms were ridiculed and discredited until competent observers recognised in the form and habits of certain aquatic animals met with in the bays and estuaries of the Indian Ocean, the Red Sea, the west coast of Africa, and sub-tropical America, the originals of these "travellers' tales." These were—first, the *manatee*, which is found in the West Indian Islands, Florida, the Gulf of Mexico, and Brazil, and in Africa in the River Congo, Senegambia, and the Mozambique Channel; second, the *dugong*, or *halicore*, which ranges along the east coast of Africa, Southern Asia, the Bornean Archipelago, and Australia; and, third, the *rytina*, seen on Behring's Island in the Kamschatkan Sea by Steller, the Russian zoologist and voyager, in 1741, and which is supposed to have become extinct within twenty-seven years after its discovery, by its having been recklessly and indiscriminately slaughtered.* Then science, in the person of Illeger, made the *amende honorable*, and frankly

* Almost all that is known of the living rytina is from an account published in 1751, in St. Petersburg, by Steller, who was one of an exploring party wrecked on Behring's Island in 1741. During the ten months the crew remained on the island they pursued this easily-captured animal so persistently, for food, that it was all but annihilated at the time. The last one there was killed in 1768.

accepting Jack's introduction to his fish-tailed *innamorata*, classed these three animals together as a sub-order of the animal kingdom, and bestowed on them the name of the *Sirenia*. This was, of course, in allusion to the Sirens of classical mythology, who, in later art, were represented as having the body of a woman above the waist, and that of a fish below, although the lower portion of their body was originally described as being in the form of a bird.

It has been found difficult to determine to which order these *Manatidæ* are most nearly allied. In shape they most closely resemble the whales and seals. But the cetacea are all carnivorous, whereas the manatee and its relatives live entirely on vegetable food. Although, therefore, Dr. J. E. Gray, following Cuvier, classed them with the cetacea in his British Museum catalogue, other anatomists, as Professor Agassiz, Professor Owen, and Dr. Murie, regard their resemblance to the whales as rather superficial than real, and conclude from their organisation and dentition that they ought either to form a group apart or be classed with the pachyderms—the hippopotamus, tapir, etc.—with which they have the nearest affinities, and to which they seem to have been more immediately linked by the now lost genera, *Dinotherium* and *Halitherium*. With the opinion of those last-named authorities I entirely agree. I regard the manatee as exhibiting a wonderful modification and adaptation of the structure of a warm-blooded land animal which enables it to pass its whole life in water, and as a connecting link between the hippopotamus, elephant, etc., on the one side, and the whales and seals on the other.

The *Halitherium* was a Sirenian with which we are only acquainted by its fossil remains found in the Miocene formation of Central and Southern Europe. These indicate that it had short hind limbs, and, consequently, approached more nearly the terrestrial type than either the manatee, the rytina, or the dugong, in which the hind limbs are absent. The two last named tend more than does the manatee to the marine mammals; but there is a strong likeness between these three recent forms. They all have a cylindrical body, like that of a seal, but instead of hind limbs there is in all a broad tail flattened horizontally; and the chief difference in their outward appearance is in the shape of this organ. In the manatee it is rounded, in the dugong forked like that of a whale, in the rytina crescent-shaped. The tail of the *Halitherium* appears to have been shaped somewhat like that of the beaver. The body of the manatee is broader in proportion to its length and depth than that of the dugong. In a paper read before the Royal Society, July 12th, 1821, on a manatee sent to London in spirits by the Duke of Manchester, then Governor of Jamaica, Sir Everard Home remarked of this greater lateral expansion that, as the manatee feeds on plants that grow at the mouths of great rivers, and the dugong upon those met with in the shallows amongst small islands in the Eastern seas, the difference of form would make the manatee more buoyant and better fitted to float in fresh water.'

In all the *Manatidæ* the mammæ of the female, which are greatly distended during the period of lactation, are situated very differently from those of the whales, being just beneath the pectoral fins. These fins or paws are much more flexible and free in their movements than those of the cetæ, and are sufficiently prehensile to enable the animal to gather food between the palms or inner surfaces of both, and the female to hold her young one to her breast with one of them. Like the whales, they are warm-blooded mammals, breathing by lungs, and are therefore obliged to come to the surface at frequent intervals for respiration. As they breathe through nostrils at the end of the muzzle, instead of, like most of the whales, through a blow-hole on the top of the head, their habit is to rise, sometimes vertically, in the water, with the head and fore part of the body exposed above the surface, and often to remain in this position for some minutes. When seen thus, with head and breast bare, and clasping its young one to its body, the female presents a certain re-

FIG. 16.—THE DUGONG. *From Sir J. Emerson Tennent's 'Ceylon.'*

semblance to a woman from the waist upward. When approached or disturbed it dives; the tail and hinder portion of the body come into view, and we see that if there was little of the "*mulier formosa superne,*" at any rate "*desinit in piscem.*" The manatee has thence been called by the Spaniards and Portuguese the "woman-fish," and by the Dutch the "manetje," or mannikin. The dugong, having the muzzle bristly, is named by the latter the "baard-manetje," or "little bearded man." There are no bristles or whiskers on the muzzle of the manatee; all the portraits of it in which these are shown are in that respect erroneous. The origin of the word "manatee" has by some been traced to the Spanish, as indicating "an animal with hands." On the west coast of Africa it is called by the natives "Ne-hoo-le." By old writers it was described as the "sea-cow." Gesner depicts it in the act of bellowing; and Mr. Bates, in his work, "The Naturalist on the Amazon," says that its voice is something like the bellowing of an ox. The Florida "crackers" or "mean whites," make the same statement. Although I have had opportunities of prolonged observation of it in captivity, I have not heard it give utterance to any sound—not even a grunt—and Mr. Bartlett, of the Zoological Gardens, tells me that his experience of it is the same. His son, Mr. Clarence Bartlett, says that a young one he had in Surinam used to make a feeble cry, or bleat, very much like the voice of a young seal. This is the only sound he ever heard from a manatee.*

I believe the dugong to be more especially the animal referred to by Ælian as the semi-human whale, and that which has led to this group having been supposed by southern voyagers to be aquatic human beings. In the first place, the dugong is a denizen of the sea, whereas the manatee is chiefly found in rivers and fresh-water lagoons; and secondly, the dugong accords with Ælian's description of the creature with a woman's face in that it has "prickles instead of hairs," whilst the manatee has no such stiff bristles.

In the case of either of these two animals being mistaken

---

* For a full description of the habits of this animal in captivity, see an article by the present writer in the 'Leisure Hour' of September 28, 1878; from which the illustration, Fig. 17, is borrowed by the kind consent of the Editor of that publication.

for a mermaid, however, "distance" must "lend enchantment to the view," and a sailor must be very impressible and imaginative who, even after having been deprived for many months of the pleasure of females' society, could be allured by the charms of a bristly-muzzled dugong, or

FIG. 17.—THE MANATEE. ITS USUAL POSITION.

mistake the snorting of a wallowing manatee for the love-song of a beauteous sea-maiden.

Unfortunately both the dugong and the manatee are being hunted to extinction.

The flesh of the manatee is considered a great delicacy.

Humboldt compares it with ham. Unlike that of the whales, which is of a deep and dark red hue, it is as white as veal, and, it is said, tastes very like it. It is remarkable for retaining its freshness much longer than other meat, which in a tropical climate generally putrefies in twenty-eight hours. It is therefore well adapted for pickling, as the salt has time to penetrate the flesh before it is tainted. The Catholic clergy of South America do not object to its being eaten on fast days, on the supposition that, with whales, seals, and other aquatic mammals, it may be liberally regarded as "fish." The "Indians" of the Amazon and Orinoco are so fond of it that they will spend many days, if necessary, in hunting for a manatee, and having killed one will cut it into slabs and slices on the spot, and cook these on stakes thrust into the ground aslant over a great fire, and heavily gorge themselves as long as the provision lasts. The milk of this animal is said to be rich and good, and the skin is valuable for its toughness, and is much in request for making leathern articles in which great strength and durability are required. The tail contains a great deal of oil, which is believed to be extremely nutritious, and has also the property of not becoming rancid. Unhappily for the dugong, its oil is in similarly high repute, and is greatly preferred as a nutrient medicine to cod-liver oil. As its flesh also is much esteemed, it is so persistently hunted on the Australian coasts that it will probably soon become extinct, like the rytina of Steller. The same fate apparently awaits the manatee, which is becoming perceptibly more and more scarce.

I fear that before many years have elapsed the Sirens of the Naturalist will have disappeared from our earth, before the advance of civilization, as completely as the fables and superstitions with which they have been connected, before the increase of knowledge ; and that the mermaid of fact will have become as much a creature of the past as the mermaid of fiction. With regard to the latter—the Siren of the poets,—the water-maiden of the pearly comb, the crystal mirror, and the sea-green tresses,—there are few persons I suppose, at the present day who would not be content to be classed with Banks, the fine old naturalist and formerly ship-mate of Captain Cook. Sir Humphry Davy in his *Salmonia* relates an anecdote of a baronet, a profound believer in these fish-tailed ladies, who on hearing some one praise very highly Sir Joseph Banks, said that " Sir Joseph was an excellent man, but he had his prejudices—he did not believe in the mermaid." I confess to having a similar " prejudice ; " and am willing to adopt the further remark of Sir Humphry Davy :—" I am too much of the school of Izaac Walton to talk of impossibility. It doubtless might please God to make a mermaid, but I don't believe God ever did make one."

THE LERNEAN HYDRA.

The mystery of the Kraken, of which I treated in a companion volume to the present, recently published, is not difficult to unravel. The clue to it is plain, and when properly taken up is as easily unwound, to arrive at the truth, as a cocoon of silk, to get at the chrysalis within it. It was a boorish exaggeration, a legend of ignorance, superstition, and wonder. But when such a skein of facts has passed through the hands of the poets, it is sure to be found in a much more intricate tangle; and many a knot of pure invention may have to be cut before it is made clear.

Nevertheless, we shall be able to discern that more than one of the most famous and hideous monsters of old classical lore originated, like the Kraken, in a knowledge by their authors of the form and habits of those strange sea-creatures, the head-footed mollusks. There can be little doubt that the octopus was the model from which the old poets and artists formed their ideas, and drew their pictures of the Lernean Hydra, whose heads grew again when cut off by Hercules; and also of the monster Scylla, who, with six heads and six long writhing necks, snatched men off the decks of passing ships and devoured them in the recesses of her gloomy cavern.

Of the Hydra Diodorus relates that it had a hundred heads; Simonides says fifty; but the generally received opinion was that of Apollodorus, Hyginus, and others, that it had only nine.

Apollodorus of Athens, son of Asclepiades, who wrote in stiff, quaint Greek about 120 B.C., gives in his 'Bibliotheca' (book ii. chapter 5, section 2) the following account of the many-headed monster. "This Hydra," he says, " nourished in the marshes of Lerne, went forth into the open country and destroyed the herds of the land. It had a huge body and nine heads, eight mortal, but the ninth immortal. Having mounted his chariot, which was driven by Iolaus, Hercules got to Lerne and stopped his horses. Finding the Hydra on a certain raised ground near the source of the Amymon, where its lair was, he made it come out by pelting it with burning missiles. He seized and stopped it, but having twisted itself round one of his feet, it struggled with him. He broke its head with his club : but that was useless ; for when one head was broken two sprang up, and a huge crab helped the Hydra by biting the foot of Hercules. This he killed, and called Iolaus, who, setting on fire part of the adjoining forest, burned with torches the germs of the growing heads, and stopped their development. Having thus out-manœuvred the growing heads, he cut off the immortal head, buried it, and put a heavy stone upon it, beside the road going from Lerne to Eleonta, and having opened the Hydra, dipped his arrows in its gall."

If we wish to find in nature the counterpart of this Hydra, we must seek, firstly, for an animal with eight outgrowths from its trunk, which it can develop afresh, or replace by new ones; in case of any or all of them being amputated or injured. We must also show that this animal, so strange in form and possessing such remarkable attributes, was well known in the locality where the legend was believed. We have it in the octopus, which abounded in the Mediterranean and Ægean seas, and whose eight prehensile arms, or tentacles, spring from its central body, the immortal head, and which, if lost or mutilated by misadventure, are capable of reproduction.

That a knowledge of the octopus existed at a very early period of man's history we have abundant evidence. The ancient Egyptians figured it amongst their hieroglyphics, and an interesting proof that they were also acquainted with other cephalopods was given to me by the late Mr. E. W. Cooke, R.A. Whilst on a trip up the Nile, in January, 1875, he visited the temple of Bayr-el-Bahree, Thebes (date 1700 B.C.), the entrance to which had been deeply buried beneath the light, wind-drifted sand, accu-

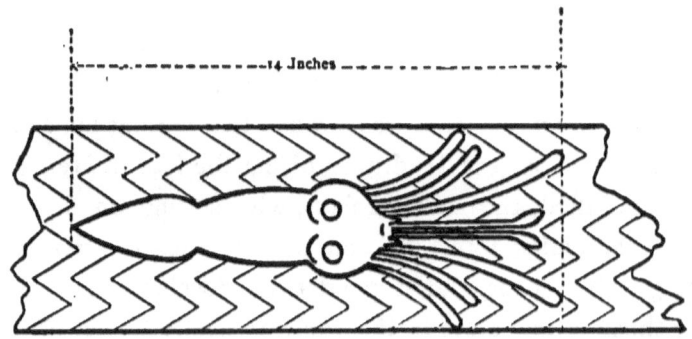

FIG. 18.—FIGURE OF A CALAMARY. *From the temple of Bayr-el-Bahree.*

mulated during many centuries. By order of the Khedive, access had just at that time been obtained to its interior, by the excavation and removal of this deep deposit, and, amongst the hieroglyphics on the walls, were found, between the zig-zag lines which represent water, figures of various fishes, copies of which Mr. Cooke kindly gave me, and which are so accurately portrayed as to be easily identified. With them was the outline of a squid fourteen inches long, a figure of which, from Mr. Cooke's drawing, is here shown. As this temple is five hundred miles from the delta of the Nile, it is remarkable that nearly all the fishes there represented are of marine species.

THE LERNEAN HYDRA.

That the octopus was a familiar object with the ancient Greeks, we know by the frequency with which its portrait is found on their coins, gems, and ornaments. Aldrovandus describes "very ancient coins" found at Syracuse and Tarentum bearing the figure of an octopus. He says the Syracusans had two coins, one of bronze, the other of gold, both of which had an octopus alone on one

FIG. 19.—FIGURE OF AN OCTOPUS ON A GOLD ORNAMENT, FOUND BY DR. SCHLIEMANN AT MYCENÆ.

side. On the reverse of the bronze one was a veiled female face in profile, with the inscription ΣΥΡΑ. I have one of these bronze Syracusan coins; it was kindly given to me, some years ago, by my friend Dr. John Millar, F.L.S. The octopus is really well depicted. On the gold coin the female head was differently veiled, and at the back of the neck was a fish. The inscription on this coin was

ΣΥΡΑΚΟΣΙΩΝ. Goltzius was of the opinion that the head was that of Arethusa. The coins found at Tarentum had on one side a figure of Neptune seated on a dolphin, and holding an octopus in one hand and a trident in the other.

Lerne, or Lerna, the reputed home of the Hydra, was a port of Southern Greece, situated at the head of the Gulf of Nauplia, and between the existing towns of Argos and Tripolitza. Within a few miles of it was Mycenæ; and it is remarkable that Dr. Schliemann, during his excavations

FIG. 20.—GOLDEN ORNAMENT IN FORM OF AN OCTOPUS, FOUND BY DR. SCHLIEMANN AT MYCENÆ.

there in 1876, found in a tomb a gold plate, or button, two and a half inches in diameter (Fig. 19), on which is figured an octopus, the eight arms of which are converted into spirals, the head and the two eyes being distinctly visible. In another sepulchre he discovered fifty-three golden models of the octopus (Fig 20), all exactly alike, and apparently cast in the same mould. The arms are very naturally carved. By the kindness of Mr. Murray, his publisher, I am enabled to give illustrations of these and two other handsome ornaments.

Having ascertained that the octopus was a familiar object in the very locality where the combat between Hercules and the Hydra is supposed to have taken place, let us compare the animal as it exists with the monstrous offspring of Typhon and Echidna.

It is a not uncommon occurrence that when an octopus is caught it is found to have one or more of its arms shorter than the rest, and showing marks of having been amputated, and of the formation of a new growth from the old cicatrix. Several such specimens were brought to the Brighton Aquarium whilst I had charge of its Natural History

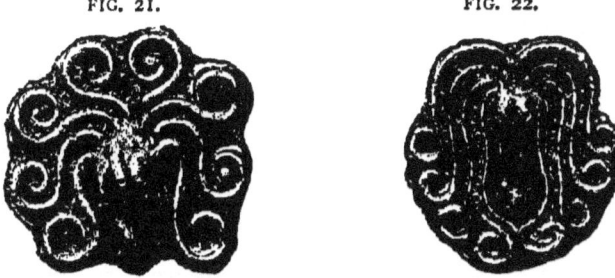

FIG. 21.　　　　　　FIG. 22.

FIGURES OF THE OCTOPUS ON GOLD ORNAMENTS FOUND BY
DR. SCHLIEMANN AT MYCENÆ.

Department. One of them was particularly interesting. Two of its arms had evidently been bitten off about four inches from the base: and out from the end of each healed stump (which in proportion to the length of the limb was as if a man's arm had been amputated halfway between the shoulder and the elbow), grew a slender little piece of newly-formed arm, about as large as a lady's stiletto, or a small button-hook—in fact just the equivalent of worthy Captain Cuttle's iron hook, which did duty for his lost hand. It was an illustrative example of the commencement of the repair and restoration of mutilated limbs.

This mutilation is so common in some localities, that

Professor Steenstrup says* that almost every octopus he has examined has had one or two arms reproduced; and that he has seen females in which all the eight arms had been lost, but were more or less restored. He also mentions a male in which this was the case as to seven of its arms. He adds that whilst the *Octopoda* possess the power of reproducing with great facility and rapidity their arms, which are exposed to so many enemies, the *Decapoda*—the *Sepiidæ* and Squids—appear to be incapable of thus repairing and replacing accidental injuries. This is entirely in accord with my own observations.

This reparative power is possessed by some other animals, of which the starfishes and crustacea are the most familiar instances. In the case of the lobster or crab, however, the only joint from which new growth can start is that connected with the body, so that if a limb be injured in any part, the whole of it must be got rid of, and the animal has, therefore, the power of casting it off at will. The octopus, on the contrary, is incapable of voluntary dismemberment, but reproduces the lost portion of an injured arm, as an out-growth from the old stump.

The ancients were well acquainted with this reparative faculty of the octopus: but of course the simple fact was insufficient for an imaginative people: and they therefore embellished it with some fancies of their own. There lingers still amongst the fishermen of the Mediterranean a very old belief that the octopus when pushed by hunger will gnaw and devour portions of its arms. Aristotle knew of this belief, and positively contradicted it; but a fallacy once planted is hard to eradicate. You may cut it down, and apparently destroy it, root and branch, but its seeds are scattered abroad, and spring up elsewhere, and in un-

* Ann. and Mag. Nat. Hist. August, 1857.

expected places. Accordingly, we find Oppian, more than five centuries later, disseminating the same old notion, and comparing this habit of the animal with that of the bear obtaining nutriment from his paws by sucking them during his hybernation.

> "When wintry skies o'er the black ocean frown,
> And clouds hang low with ripen'd storms o'ergrown,
> Close in the shelter of some vaulted cave
> The soft-skinn'd prekes* their porous bodies save.
> But forc'd by want, while rougher seas they dread,
> On their own feet, necessitous, are fed.
> But when returning spring serenes the skies,
> Nature the growing parts anew supplies.
> Again on breezy sands the roamers creep,
> Twine to the rocks, or paddle in the deep.
> Doubtless the God whose will commands the seas,
> Whom liquid worlds and wat'ry natives please,
> Has taught the fish by tedious wants opprest
> Life to preserve and be himself the feast.

The fact is, that the larger predatory fishes regard an octopus as very acceptable food, and there is no better bait for many of them than a portion of one of its arms: Some of the cetacea also are very fond of them, and whalers have often reported that when a "fish" (as they call it) is struck it disgorges the contents of its stomach, amongst which they have noticed parts of the arms of cuttles which, judging from the size of their limbs, must have been very large specimens. The food of the sperm whale consists largely of the gregarious squids, and the presence in spermaceti of their undigested beaks is accepted as a test of its being genuine. That old fish-

---

* The octopus is still called the "preke" in some parts of England, notably in Sussex. The translation of Oppian's 'Halieutics,' from which this passage and others are quoted is that by Messrs. Jones and Diaper, of Baliol College, Oxford, and was published in 1722.

reptile, the Ichthyosaurus, also, preyed upon them; and portions of the horny rings of their suckers were discovered in its coprolites by Dean Buckland. Amongst the worst enemies of the octopus is the conger. They are both rock-dwellers, and if the voracious fish come upon his cephalopod neighbour unseen, he makes a meal of him, or, failing to drag him from his hold, bites off as much of one or two of his arms as he can conveniently obtain. The conger, therefore, is generally the author of the injury which the octopus has been unfairly accused of inflicting on itself.

Continuing our comparison with the hydra, we have in the octopus an animal capable of quitting its rocky lurking-place in the sea, and going on a buccaneering expedition on dry land. Many incidents have been related in connection with this; but I can attest it from my own observation. I have seen an octopus travel over the floor of a room at a very fair rate of speed, toppling and sprawling along in its own ungainly fashion; and in May, 1873, we had one at the Brighton Aquarium which used regularly every night to quit its tank, and make its way along the wall to another tank at some distance from it, in which were some young lump-fishes. Day after day, one of these was missing, until, at last, the marauder was discovered. Many days elapsed, however, before he was detected, for after helping himself to, and devouring a young "lump-sucker," he demurely returned before daylight to his own quarters.

Of this habit of the octopus the ancients were, also, fully aware. Aristotle wrote that it left the water and walked in stony places, and Pliny and Ælian related tales of this animal stealing barrels of salt fish from the wharves, and crushing their staves to get at the contents. An octopus that could do this would be as formidable a predatory monster as the Lernean Hydra, which had the evil reputation of devouring the Peloponnesian cattle.

Whoever first described the counter-attack of the Hydra on Hercules must have had the octopus in his thoughts. "It twisted itself round one of his feet"—exactly that which an octopus would do.

Finally, according to the legend, Hercules dipped his arrow-heads in the gall of the Hydra, and, from its poisonous nature, all the wounds he inflicted with them upon his

FIG. 23.—HERCULES SLAYING THE LERNEAN HYDRA.
From Smith's 'Classical Dictionary.'

enemies proved fatal. It is worthy of notice that the ancients attributed to the octopus the possession of a similarly venomous secretion. Thus Oppian writes:

"The crawling preke a deadly juice contains
Injected poison fires the wounded veins."

The accompanying illustration (Fig. 23) of Hercules slaying the Hydra is taken from a marble tablet in the Vatican. It will be immediately seen how closely the Hydra, as there depicted, resembles an octopus. The body is elongated, but the eight necks with small heads on them bear about the same proportion to the body as the arms to the body of an octopus.

The Reverend James Spence, in his 'Polymetis,' published in 1755, gives a figure, almost the counterpart of this, copied from an antique gem, a carnelian, in the collection of the Grand Duke of Tuscany at Florence. Only seven necks of the hydra are, however, there visible, and there are two coils in the elongated body. On the upper part are two spots which have been supposed to represent breasts. This was probably intended by the artificer; but that the idea originated from a duplication of the syphon tube is evident from the figures (Figs. 21, 22) of the octopus on the smaller gold ornaments found by Dr. Schliemann at Mycenæ. In the same work is also an engraving from a picture in the Vatican Virgil, entitled 'The River, or Hateful Passage into the Kingdom of Ades,' wherein an octopus—hydra, of which only six heads and necks are shown, is one of the monsters called by the author "Terrors of the Imagination."

SCYLLA AND CHARYBDIS.

IN the description given by Homer, in the twelfth book of the 'Odyssey,' of the unfortunate nymph Scylla, transformed by the arts of Circe into a frightful monster, the same typical idea as in the case of the Hydra is perceptible. The lurking octopus, having its lair in the cranny of a rock, watching in ambush for passing prey, seizing anything coming within its reach with one or more of its prehensile arms, even brandishing these fear-inspiring weapons out of water in a threatening manner, and known in some localities to be dangerous to boats and their occupants, is transformed into a many-headed sea monster, seizing in its mouths, instead of by the adhesive suckers of its numerous arms, the helpless sailors from passing vessels, and devouring them in the abysses of its cavernous den.

Circe, prophesying to Ulysses the dangers he had still to encounter, warned him especially of Scylla and Charybdis, within the power of one of whom he must fall in passing through the narrow strait (between Italy and Sicily) where they had their horrid abode. Describing the lofty rock of Scylla, she tells him:

> "Full in the centre of this rock displayed
> A yawning cavern casts a dreadful shade,
> Nor the fleet arrow from the twanging bow
> Sent with full force, could reach the depth below.
> Wide to the west the horrid gulf extends,
> And the dire passage down to hell descends.

> O fly the dreadful sight! expand thy sails,
> Ply the strong oar, and catch the nimble gales;
> Here Scylla bellows from her dire abodes;
> Tremendous pest! abhorred by man and gods!
> Hideous her voice, and with less terrors roar
> The whelps of lions in the midnight hour.
> Twelve feet deformed and foul the fiend dispreads;
> Six horrid necks she rears, and six terrific heads;
> * . * * * * *
> When stung with hunger she embroils the flood,
> The sea-dog and the dolphin are her food;
> She makes the huge leviathan her prey,
> And all the monsters of the wat'ry way;
> The swiftest racer of the azure plain
> Here fills her sails and spreads her oars in vain;
> Fell Scylla rises, in her fury roars,
> At once six mouths expands, at once six men devours." *

Circe then describes the perils of the whirling waters of Charybdis as still more dreadful; and, admonishing Ulysses that once in her power all must perish, she advises him to choose the lesser of the two evils, and to

> "shun the horrid gulf, by Scylla fly;
> 'Tis better six to lose than all to die."

Ulysses continues his voyage; and as his ship enters the ominous strait,

> "Struck with despair, with trembling hearts we viewed
> The yawning dungeon, and the tumbling flood;
> When, lo! fierce Scylla stooped to seize her prey,
> Stretched her dire jaws, and swept six men away.
> Chiefs of renown! loud echoing shrieks arise;
> I turn, and view them quivering in the skies;
> They call, and aid, with outstretched arms, implore,
> In vain they call! those arms are stretched no more.
> As from some rock that overhangs the flood,
> The silent fisher casts th' insidious food;

---

* Homer's 'Odyssey,' Pope's Translation, Book XII.

With fraudful care he waits the finny prize,
And sudden lifts it quivering to the skies;
So the foul monster lifts her prey on high,
So pant the wretches, struggling in the sky;
In the wide dungeon she devours her food,
And the flesh trembles while she churns the blood."

THE "SPOUTING" OF WHALES.

ONE of the sea-fallacies still generally believed, and accepted as true, is that whales take in water by the mouth, and eject it from the spiracle, or blow-hole.

The popular ideas on this subject are still those which existed hundreds of years ago, and which are expressed by Oppian in two passages in his 'Halieutics':

> "Uncouth the sight when they in dreadful play
> Discharge their nostrils and refund a sea,"

and

> "While noisy fin-fish let their fountains fly
> And spout the curling torrent to the sky."

Eminent zoologists and intelligent observers, who have had full opportunities of obtaining practical knowledge of the habits of these great marine mammals, have forcibly combated and repeatedly contradicted this erroneous idea; but their sensible remarks have been read by few, in comparison with the numbers of those to whom a wrong impression has been conveyed by sensational pictures in which whales are represented *with their heads above the surface*, and throwing up from their nostrils columns of water, like the fountains in Trafalgar Square. One can hardly be surprised that the old writers on Natural History were unacquainted with the real composition of the whale's "spout." Those of them who sought for any original information on marine zoology, obtained it chiefly from uninstructed and superstitious fishermen; but they generally contented themselves with diligent compilation, and thus copied and transmitted the errors of their predecessors, with the addition of some slight embellishments of their own. Accordingly, we find Olaus Magnus * describing, as follows, the *Physeter*, or, as his translator, Streater, calls it, the *Whirlpool*. "The *Physeter* or *Pristis*," he says, "is a kind of whale, two hundred cubits long, and is very cruel. For, to the danger of seamen, he will sometimes raise himself above the sail-yards, and casts such floods of waters above his head, which he had sucked in, that with a cloud of them he will often sink the strongest ships, or expose the mariners to extreme danger. This beast hath also a large round mouth, like a lamprey, whereby he sucks in his meat or water, and by his weight cast upon the fore or hinder deck, he sinks and drowns a ship."

Figures 24 and 25 (p. 64) are facsimiles of the illustrations which accompany the above description. It will be seen that, in the first, the *Physeter* is depicted as uprearing a maned neck and head, like that of a fabled dragon; whilst in Fig. 25 it is shown as a whale flinging itself on board a ship, which is sinking under its ponderous weight. In both, torrents of water are issuing from its head, and it is evident that they are merely exaggerated misrepresentations of the "spouting" of whales.

Gesner copies many of Olaus Magnus's illustrations, and improves upon Fig. 25 by putting a numerous crew on board the ship. The unfortunate sailors are depicted in every attitude of terror and despair, and seem to be incapacitated from any attempt to save themselves by the flood of water which the whale is deliberately pouring upon them from its blow-holes.

* 'Historia de Gentibus Septentrionalibus,' lib. xxi. cap. vi. A.D. 1555.

FIG. 24.—THE PHYSETER INUNDATING A SHIP. *After Olaus Magnus.*

These old pictures appear, no doubt, ridiculous, but they are, really, very little more absurd and untrue to nature than many of those which disfigure some otherwise useful books on Natural History of the present day. I could

FIG. 25.—A WHALE POURING WATER INTO A SHIP FROM ITS BLOW-HOLE. *After Olaus Magnus.*

refer to several, in which whales are represented as spouting from their blow-holes one or more columns of water, which, after ascending skyward to a considerable distance, fall

FIG. 26.—SPERM WHALES SPOUTING.

over gracefully as if issuing from the nozzle of an ornamental fountain. I select one from amongst them (Fig. 26), not with any disrespect for the artist, author, or publisher of the work from which it is taken, but because, whilst it shows correctly the position of the blow-hole of the sperm whale, it also exhibits exactly that which I wish to confute. The publishers of the valuable work in which this picture appeared have generously consented to my reproducing it here.

When, in describing, in 1877, the White Whale then exhibited at the Westminster Aquarium, I said that whales do not spout water out of their blow-holes, and that the idea that they do so is a popular error, the statement was so contrary to generally-accepted notions that I was not surprised by receiving more than one letter on the subject. One very reasonable suggestion made to me was that, although the lesser whales, such as the porpoises, which I had had opportunities of watching in confinement at Brighton for two years, and the *Beluga*, which had been observed for a similar period at the New York Aquarium, and also at Westminster, did not "spout," the respiratory apparatus of the larger whales might be so modified as to permit them to do so. Let us consider the construction of the breathing apparatus which would have to be thus modified, as shown in the porpoise.

In the first place, there is a pair of lungs as perfect as those of any land mammal, fitted to receive air, and to bring the hot blood into contact with the air, that it may absorb the oxygen of the air, and so be purified. But this air cannot well be breathed through the mouth of an animal which has to take its food from and in water; so it has to be inhaled only by the nostrils. If these were situated as they are in land mammals, near the extremity of the nose, the porpoise would be obliged to stop when pursuing its prey, or, escaping from its enemies, to put the tip of its nose above the surface of the water every time it required to breathe. A much more convenient arrangement has, therefore, been provided for it, and for almost all whales, by which that difficulty is removed. Instead of running along the bones of the nose, the nostrils are placed on the top of the head, and the windpipe is turned up to them without having any connection with the palate. The upper jaw is quite solid. Thus the mouth is solely devoted to the reception of food, and the animal is enabled to continue its course when swimming, however rapidly, by rising obliquely to the surface, and exposing the top of its head above it. On the blow-hole being opened, the air, from which the oxygen has been absorbed, is expelled in a sudden puff, another supply is instantaneously inhaled, and rushes into the lungs with extreme velocity, and then the porpoise can either descend into the depths, or remain with its spiracle exposed to the air, as it may prefer. In this act of breathing the spiracle is normally brought above the water, the breath escapes, and the immediate inhalation is effected almost in silence. But frequently, and in some whales habitually, the blow-hole is opened just below the surface, and then the outrush of air causes a splash upwards of the water overlying it.

I may here mention that I have frequently seen the porpoises at the Brighton Aquarium lying asleep at the surface, with the blow-hole exposed above it, breathing automatically, and without conscious effort. Aristotle was acquainted with this habit of the cetacea 2,200 years ago, for he wrote : " They sleep with the blow-hole, their organ of respiration, elevated above the water."

The apparatus for closing the blow-hole, so that not a drop of water shall enter the windpipe, even under great pressure, is a beautiful contrivance, complex in its structure, yet most simple in its working. The external aperture is covered by a continuation of the skin, locally thickened, and connected with a conical stopper, of a texture as tough as india-rubber, which fits perfectly into a cone or funnel formed by the extremity of the windpipe, and closes more and more firmly as the pressure upon it is increased. Whilst the orifice is thus guarded, the lower end of the tube is surrounded by a strong compressing muscle, which clasps also the glottis, and thus the passage from the blow-hole to the lungs is completely stopped.

There is nothing in this which indicates the possibility of the spouting of water from the nostrils; but as assertions that water had been seen to issue from them were positive and persistent, anatomists seem to have felt themselves obliged to try to account for it somehow. Accordingly the theory was propounded by F. Cuvier that the water taken into the mouth is reserved in two pouches (one on each side), until the whale rises to blow, when, the gullet being closed, it is forced by the action of the tongue and jaws through the nasal passages, somewhat as a smoker occasionally expels the smoke of his cigar through his nostrils. Although these pouches, or sacs analogous to them, are found at the base of the nostrils of the horse, tapir, etc.,—animals which do not "spout" from the nostrils water taken in by the mouth—the explanation was accepted for a time.

Mr. Bell held this opinion when the first edition of his 'British Quadrupeds' was published in 1837, but before the issue of the second edition, in 1874, he had found reasons for taking a different view of the matter; and, under the advice of his judicious editors, Mr. Alston, and Professor Flower (the latter of whom supervised the proofs of the chapters on the Cetacea) his sanction of the illusion was withdrawn as follows :—" The results of more recent and careful observations, amongst which we may notice those of Bennett, Von Baer, Sars and Burmeister, are directly opposed to the statement that water is thus ejected ; and there can now be no doubt that the appearance which has given rise to the idea is caused by the moisture with which the expelled breath is supercharged, which condenses at once in the cold outer air, and forms a cloud or column of white vapour. It is possible indeed that if the animal begins to 'blow' before its head is actually at the surface, the force of the rushing air may drive up some little spray along with it, but this is quite different from the notion that water is really expelled from the nasal passages. We may add that on the only occasion when we ourselves witnessed the 'spouting' of a large whale we were much struck with its resemblance to the column of white spray which is dashed up by the ricochetting ball fired from one of the great guns of a man-of-war."

The simile is admirable, and nothing could better describe the appearance of a whale's "spout"; but, in the previous portion of the passage (except with reference to the sperm whale, the nostrils of which are not on the top of the head), I think sufficient importance is not conceded to the volume of water propelled into the air by the outrush of breath from the submerged blow-hole. I do not know how many cubic feet of air the lungs of a great whale are capable of containing, but the quantity is sufficient to force up to a height of several feet the water above the valve when the latter is opened, not only in "some little spray," but, for some distance in a good solid jet—enough, in fact, to give the appearance of its actually issuing from the blow-hole, and to account for the erroneous belief of sailors that it does so. It must be remembered that the escape of air is not by a prolonged wheeze, but by a sudden blast, and thus when the spiracle is opened just beneath the surface, an instant before it is uncovered to take in a fresh supply of air, the water above its orifice is thrown up as by a slight subaqueous explosion, or as by the momentary opening under water of the safety-valve of a steam boiler. Some idea of the force and volume of the blast of air from the lungs of even the common porpoise may be formed when I mention that one of the porpoises at the Brighton Aquarium, happening to open its spiracle just beneath an illuminating gas jet fixed over its tank, blew out the light.

In the sperm whale the nostrils are placed near the extremity of the nose, and therefore this whale has to raise its snout above the surface when it requires to breathe; but instead of this being necessary, as in the case of the porpoise twice or thrice in a minute, the sperm whale only rises to "blow" at intervals of from an hour to an hour and twenty minutes. Mr. Beale says[*] that in a large bull sperm whale the time consumed in making one expiration and one inspiration is ten seconds, during six of which the nostril is beneath the surface of the water—the expiration occupying three seconds, and the inspiration one second. At each breathing time this whale makes from sixty to seventy expirations, and remains, therefore, at the surface ten or eleven minutes, and then, raising its tail, it descends perpendicularly, head first. In different individuals the time required for performing these several acts varies; but in each they are minutely regular, and this well-known regularity is of considerable use to the fishers, for when a whaler has once noticed the periods of any particular whale which is not alarmed, he knows to a minute when to expect it to come to the surface, and how long it will remain there. The "spout" of the sperm whale differs much from that of other whales. Unlike, for instance, the straight perpen-

[*] 'Natural History of the Sperm Whale.' Van Voorst, 1839.

dicular twin jets of the "right whale," the single, forward-slanting "spout" of the sperm whale presents a thick curled bush of white mist.. Each whale has a different mode and time of breathing, and the form of the "spout" differs accordingly.

It is said that the blowing of the *Beluga*, or "White Whale," is not unmusical at sea, and that when it takes place under water it often makes a peculiar sound which might be mistaken for the whistling of a bird., Hence is derived one of the names given to this whale by sailors—the "Sea-canary." Though I have had opportunities of attentively watching the breathing and other actions in captivity of two specimens of this whale I have never been able to detect the sound alluded to.

Besides the opinions cited by Mr. Bell concerning whales spouting water from their blow-holes, we have other evidence which is most clear and definite, and which ought to be convincing.

We will take first that of Mr. Beale, who as surgeon on board the "Kent" and "Sarah and Elizabeth," South Sea whalers, passed several seasons amongst sperm whales. He says:—"I can truly say when I find myself in opposition to these old and received notions, that out of the thousands of sperm whales which I have seen during my wanderings in the South and North Pacific Oceans, I have never observed one of them to eject a column of water from the nostril. I have seen them at a distance, and I have been within a few yards of several hundreds of them, and I never saw water pass from the spout-hole. But the column of thick and dense vapour which is certainly ejected is exceedingly likely to mislead the judgment of the casual observer in these matters; and this column does indeed appear very much like a jet of water when seen at the distance of one or two miles on a clear day, because of the condensation of the vapour which takes place the moment it escapes from the nostril, and its consequent opacity, which makes it appear of a white colour, and which is not observed when the whale is close to the spectator. It then appears only like a jet of white steam. The only water in addition is the small quantity that may be lodged in the external fissure of the spout hole, when the animal raises it above the surface to breathe, and which is blown up into the air with the 'spout,' and may probably assist in condensing the vapour of which it is formed. . . . I have been also very close to the *Balæna mysticetus* (the Greenland, or Right whale) when it has been feeding and breathing, and yet I never saw even that animal differ in the latter respect from the sperm whale in the nature of the spout. . . . If the weather is fine and clear, and there is a gentle breeze at the time, the spout may be seen from the masthead of a moderate-sized vessel at the distance of four or five miles."

Captain Scoresby, who was a veteran and successful whaler, a good zoologist, and a highly intelligent observer, says :—" A moist vapour mixed with mucus is discharged from the nostrils when the animal breathes ; but no water accompanies it unless an expiration of the breath be made under the surface."

Dr. Robert Brown, who communicated to the Zoological Society, in May, 1868, a valuable series of observations on the mammals of Greenland, made during his voyages to the Spitzbergen, Iceland, and Jan Mayen Seas, and along the eastern and western shores of Davis's Strait and Baffin's Bay to near the mouth of Smith's Sound, remarks, in a chapter on the Right whale (*Balæna mysticetus*) :—" The 'blowing,' so familiar a feature of the *Cetacea*, but especially of the *Mysticetus* is, quite analogous to the breathing of the higher mammals, and the blow-holes are the homologues of the nostrils. It is most erroneously stated that the whale ejects water from the blow-holes. I have been many times only a few feet from a whale when 'blowing,' and, though purposely observing it, could never see that it ejected from its nostrils anything but the ordinary breath—a fact which might almost have been deduced from analogy. In the cold arctic air this breath is generally condensed, and falls upon those close at hand in the form of a dense spray which may have led seamen to suppose that this vapour was originally ejected in the form of water. Occasionally, when the whale blows just as it is rising out of or sinking in the sea, a little of the superincumbent water may be forced upwards by the column of breath. When the whale is wounded in the lungs, or in any of the blood-vessels immediately supplying them, blood, as might be expected, is ejected in the death-throes along with the breath. When the whaleman sees his prey 'spouting red,' he concludes that its end is not far distant; it is then mortally wounded."

Captain F. C. Hall, the commander of the unfortunate "Polaris" Expedition, thus describes, in his 'Life with the Esquimaux,' the spout of a whale :—"What this blowing is like," he says, "may be described by asking if the reader has ever seen the smoke produced by the firing of an old-fashioned flint-lock. If so, then he may understand the 'blow' of a whale—a flash in the pan and all is over."

Captain Scammon, an experienced American whaling captain, who, like Scoresby, could wield well both harpoon and pen, in his fine work on 'The Marine Mammals of the North-Western Coast of America,' writes to the same effect.

Mr. Herman Melville, who is not a naturalist, but has served before the mast in a sperm-whaler and borne his part in all the hardships and dangers of the chase, writes, in his remarkable book, 'The Whale':—'As for this 'whale-spout' you might almost stand in it, and yet be undecided as to what it is precisely. Nor is it at all prudent for the hunter to be over curious respecting it. For, even when coming into slight contact with the outer vapoury shreds of the jet, which will often happen, your skin will feverishly smart from the acrimony of the thing so touching you. And I know one who, coming into still closer contact with the spout—whether with some scientific object in view or otherwise I cannot say—the skin peeled off from his cheek and arm. Wherefore, among whalemen, the spout is deemed poisonous; they try to evade it. I have heard it said, and I do not much doubt it, that if the jet were fairly spouted into your eyes it would blind you."

The only other eye-witness I will cite is Mr. Bartlett, of the Zoological Gardens, whose experience and accuracy as an observer of the habits of animals is unsurpassed. He spent an autumn holiday in accompanying the late Mr. Frank Buckland and his colleagues, Messrs. Walpole and Young, in a tour of inquiry into the condition of the herring fishery in Scotland. When the commissioners left Peterhead, he remained there for a few days as the guest of Captain David Gray, of the steam whaler, "Eclipse," and as it was reported that large whales had been seen in the offing, his host invited him to go in search of them, and pay them a visit in his steam-launch. When about twelve miles out, they saw the whales, which were "finners," at a distance of four or five miles. Fourteen were counted—all large ones—some of which were seventy feet in length. On approaching them the captain shut off steam, and the launch was allowed to float in amongst them. So close were they to the boat that it would not have been difficult to jump upon the back of one of them had that been desirable. Mr. Bartlett tells me that he was greatly astonished by the immense force of the sudden outrush of air from their blow-holes, and the noise by which it was accompanied. He believes that the blast was strong enough to blow a man off the spiracle if he were seated on it. He authorizes me to say that having seen and watched these whales under such favourable circumstances, he entirely agrees with all that I have here written concerning the so-called "spout." The volume of hot, vaporous breath expelled is enormous, and this is accompanied by no small quantity of water, forced up by it when the blow-hole is opened below the surface.

An effect similar in appearance to the whale's spout is produced by the breathing of the hippopotamus. When this great beast opens its nostrils beneath the surface, water and spray are driven and scattered upward by the force of the air, but, of course, do not issue from the nasal passages. I have, also, seen this effect produced, though in a less degree, by the breathing of sea-lions.

I repeat, therefore, that not a drop of sea-water enters or passes out of the blow-hole of a whale. If the spiracle valve were in a condition to allow it to do so the animal would soon be drowned. Everyone knows the extreme irritation and the horrible feeling of suffocation caused to a human being, whilst eating or drinking, by a crumb or a little liquid "going the wrong way"—that is, being accidentally drawn to the air-passages instead of passing to the œsophagus. If water were to enter the bronchi of a whale it would instantly produce similar discomfort.

The neck of a popular error is hard to break; but it is time that one so palpable as that concerning the "spouting" of whales should cease to be promulgated and disseminated by fanciful illustrations of instructive books.

THE "SAILING" OF THE NAUTILUS.

ONE of the prettiest fables of the sea is that relating to the Paper Nautilus, the constructor and inhabitant of the delicate and beautiful shell which looks as if it were made of ivory no thicker than a sheet of writing paper.

FIG. 27.—THE PAPER NAUTILUS (*Argonauta argo*) SAILING.

It is an old belief that in calm weather it rises from the bottom of the sea, and, elevating its two broadly-expanded arms, spreads to the gentle air, as a sail, the membrane, light as a spider's web, by which they are united; and that, seated in its boat-like shell, it thus floats over the smooth surface of the ocean, steering and paddling with its other arms. Should storm arise or danger threaten, its masts and sail are lowered, its oars laid in, and the frail craft, filling with water, sinks gently beneath the waves.

When and where this picturesque idea originated I am unable to discover. It dates far back beyond the range of history; for Aristotle mentions it, and, unfortunately, sanctioned it. With the weight of his honoured name in its favour, this fallacy has maintained its place in popular belief, even to our own times; for the mantle of the great father of natural history, who was generally so marvellously correct, fell on none of his successors; Pliny, and Ælian, and the tribe of compilers who succeeded them, having been more concerned to make their histories sensational than to verify their statements.

Naturally, the Paper Nautilus has been the subject of many a poet's verses. Oppian wrote of it in his 'Halieutics':—

> "Sail-fish in secret, silent deeps reside,
> In shape and nature to the preke * allied;
> Close in their concave shells their bodies wrap,
> Avoid the waves and every storm escape.
> But not to mirksome depths alone confined;
> When pleasing calms have stilled the sighing wind,
> Curious to know what seas above contain,
> They leave the dark recesses of the main;
> Now, wanton, to the changing surface haste,
> View clearer skies, and the pure welkin taste.
> But slow they, cautious, rise, and, prudent, fear
> The upper region of the watery sphere;
> Backward they mount, and as the stream o'erflows,
> Their convex shells to pressing floods oppose.
> Conscious, they know that, should they forward move,
> O'erwhelming waves would sink them from above,

---

* The octopus.

Fill the void space, and with the rushing weight,
Force down th' inconstants to their former seat.
When, first arrived, they feel the stronger blast,
They lie supine and skim the liquid waste.
The natural barks out-do all human art
When skilful floaters play the sailor's part.
Two feet they upward raise, and steady keep;
These are the masts and rigging of the ship:
A membrane stretch'd between supplies the sail,
Bends from the masts, and swells before the gale.
Two other feet hang paddling on each side,
And serve for oars to row and helm to guide.
'Tis thus they sail, pleased with the wanton game,
The fish, the sailor, and the ship, the same.
But when the swimmers dread some dangers near
The sportive pleasure yields to stronger fear.
No more they, wanton, drive before the blasts,
But strike the sails, and bring down all the masts;
The rolling waves their sinking shells o'erflow,
And dash them down again to sands below."

Montgomery also thus exquisitely paraphrases the same idea in his 'Pelican Island':—

"Light as a flake of foam upon the wind,
Keel upwards, from the deep emerged a shell,
Shaped like the moon ere half her orb is filled.
Fraught with young life, it righted as it rose,
And moved at will along the yielding water.
The native pilot of this little bark
Put out a tier of oars on either side,
Spread to the wafting breeze a twofold sail,
And mounted up, and glided down, the billows
In happy freedom, pleased to feel the air,
And wander in the luxury of light."

Byron mentions the Nautilus in his 'Mutiny of the Bounty' as follows:—

"The tender Nautilus, who steers his prow,
The sea-born sailor of his shell canoe,
The ocean Mab—the fairy of the sea,
Seems far less fragile, and alas! more free.

He, when the lightning-winged tornadoes sweep
The surge, is safe: his port is in the deep;
And triumphs o'er the armadas of mankind
Which shake the world, yet crumble in the wind."

The very names by which this animal is known to the science which some persons erroneously think must be so hard and dry are poetic. In Aristotle's day it was called the *Nautilus* or *Nauticus*, "the mariner," and though two thousand two hundred years have passed since the great master wrote, the name still clings to it. As the Pearly Nautilus, a very different animal, also bears that name, Gualtieri perceived the necessity of distinguishing the Paper Nautilus from it, and was followed by Linnæus, who therefore entitled the genus to which the latter belongs, *Argonauta*, after the ship *Argo*, in which Jason and his companions sailed to Colchis to carry off the "Golden Fleece" suspended there in the temple of Mars, and guarded by brazen-hoofed bulls, whose nostrils breathed out fire and death, and by a watchful dragon that never slept. According to the Greek legend, the *Argo* was named after its builder Argus, the son of Danaus, and was the first ship that ever was built. Oppian ('Halieutics,' book I.) expresses his opinion that the Nautilus served as a model for the man who first conceived the idea of constructing a ship, and embarking on the waters :—

"Ye Powers! when man first felled the stately trees,
And passed to distant shores on wafting seas,
Whether some god inspired the wondrous thought,
Or chance found out, or careful study sought;
If humble guess may probably divine,
And trace th' improvement to the first design,
Some wight of prying search, who wond'ring stood
When softer gales had smoothed the dimpled flood,
Observed these careless swimmers floating move,
And how each blast the easy sailor drove;

Hence took the hint, hence formed th' imperfect draught,
And ship-like fish the future seaman taught.
Then mortals tried the shelving hull to slope,
To raise the mast, and twist the stronger rope,
To fix the yards, let fly the crowded sails,
Sweep through the curling waves, and court auspicious gales."

Pope, too, in his 'Essay on Man' (Ep. 3), adopted the idea in his exhortation—

"Learn of the little Nautilus to sail,
Spread the thin oar, and catch the driving gale."

Poetry, like the wizard's spell, can make

"A nutshell seem a gilded barge,
A sheeling seem a palace large,"

but the equally enchanting wand of science is able by a touch to dispel the illusion, and cause the object to appear in its true proportions. So with the fiction of the "Paper Sailor."

I have elsewhere described the affinities of the Nautili and their place in nature, therefore it will only be necessary for me here to allude to these very briefly, to explain the great and essential difference that exists between the two kinds of Nautilus which are popularly regarded as being one and the same animal.

The *Pearly* Nautilus (*Nautilus pompilius*) and the Argonaut, which from having a fragile shell of somewhat similar external form is called the *Paper* Nautilus, both belong to that great primary group of animals known as the *Mollusca*, and to the class of it called the *Cephalopoda*, from their having their head in the middle of that which is the foot in other mollusks. In the Cephalopoda the foot is split or divided into eight segments in some families, and in others into ten segments, which radiate from the central head, like so many rays. These rays are not only used as feet, but, being highly flexible, are adapted for employment also as prehensile arms, with which their owner captures its prey, and they are rendered more perfect for this purpose by being furnished with suckers which hold firmly to any surface to which they are applied. The Cephalopods which have the foot divided into ten of these segments or arms are called the *Decapoda*, those which have only eight of them are called the *Octopoda*. All of these have *two* plume-like gills—one on each side—and so are called *Dibranchiata*; and in the eight-armed section of these is the argonaut or Paper Nautilus. Of the Pearly Nautilus and the four-gilled order I shall have more to say by-and-by: at present we will follow the history of the argonaut.

Notwithstanding all that has been written of it, it is only within the last fifty years that this has been correctly understood. An eight-armed cuttle was recognised and named *Ocythoe*, which, instead of having, like the common octopus, all of its eight arms thong-like and tapering to a point, had the two dorsal limbs flattened into a broad thin membrane.

FIG. 28.—THE PAPER NAUTILUS (*Argonauta argo*) RETRACTED WITHIN ITS SHELL.

Although this animal was sometimes seen dead without any covering, it was generally found contained in a thin and slightly elastic univalve shell of graceful form, and bearing some resemblance to an elegantly shaped boat. It did not penetrate to the bottom of this shell; it was not attached to it by any muscular ligament, nor was the shell moulded on its body, nor apparently made to fit it. Hence it was long regarded as doubtful, and even by naturalists so recent and eminent as Dumeril and De Blainville, whether the octopod really secreted the shell, or whether, like the hermit-crab, it borrowed for its protection the shell of some other mollusk. Aristotle left the subject with the faithful acknowledgment: "As to the origin and growth of this shell nothing is yet exactly determined. It appears to be produced like other shells; but even this is not evident, any more than it is whether the animal can live without it." Pliny, as usual, instead of throwing light on the matter, obscured it. He regarded the shell as the property of a gasteropod like the snail, and the octopod as an amateur yachtsman who occasionally went on board and took a trip in the frail craft, and assisted its owner to navigate it for the fun of the thing. This is what he says about it*: "Mutianus reports that he saw in the Propontis a shell formed like a little ship, having the poop turned up and the prow pointed. An animal called the *Nauplius*, resembling an octopus, was enclosed in the shell with its owner, for its amusement in the following manner. When the sea is calm the guest lowers his arms, and uses them as oars and a helm, whilst the owner of the shell expands himself to catch the wind; so that one has the pleasure of carrying and sailing, and the other of steering. Thus, these two otherwise senseless animals take their pleasure together; but the meeting them sailing in their shell is a bad omen for mariners, and foretells some great calamity."

Although the animal was never found in any other shell, and the shell was never known to contain any other animal, and though, when the shell and the animal were found together they were always of proportionate size, this octopod, as I have said, was looked upon by some conchologists as a pirate who had taken possession of a ship which did not belong to him, until Madame Jeannette Power, a French lady then

* Naturalis Historia, lib. ix. cap. 30.

residing in Messina, having succeeded in keeping alive for a time an argonaut the shell of which had been broken in its capture, discovered that the animal quickly repaired the fracture, and reproduced the portions that had been broken off. Induced by this to make further experiments, she kept a number of living argonauts in cages sunk in the sea near the citadel of Messina, and in 1836 laid before the "Academy" at Catania the following results of her observations of them :—

1st. That the argonaut constructs the shell which it inhabits.

2nd. That it quits the egg entirely naked, and forms the shell after its birth.

3rd. That it can repair its shell, if necessary, by a fresh deposit of material having the same chemical composition as its original shell.

4th. That this material is secreted by the palmate, or sail, arms, and is laid on the outside of the shell, to the exterior of which these membranous arms are closely applied.

Madame Power was mistaken on two points. Firstly, the construction of the shell does not commence after the birth of the animal, but, as has been shown by M. Duvernoy, its rudimentary form is distinctly visible by the aid of the microscope in the embryo, whilst still in the egg; and secondly, she continued to believe in the use of the membranous arms as sails, and of the others as oars. This fallacy was exploded by Captain Sander Rang, an officer of the French navy, and "port-captain" at Algiers, who carefully followed up Madame Power's experiments, and confirmed the more important of them. Thus were set at rest questions which for centuries had divided the opinions of zoologists.

The "Paper Nautilus" is, in fact, a female octopod provided with a portable nest, in which to carry about and protect her eggs, instead of brooding over them in some cranny of a rock, or within the recesses of a pile of shells, as does her cousin the octopus. From the membranes of the two flattened and expanded arms she secretes and, if necessary, repairs her shell, and by applying them closely to its outer surface on each side, holds herself within it, for it is not fastened to her body by any attaching muscles. When disturbed or in danger she can loosen her hold, and, leaving her cradle, swim away independently of it. It has been said that, having once left it, she has not the ability nor perhaps the sagacity to re-enter her nest, and resume the guardianship of her eggs."* From my own observations of the breeding habits of other octopods I think this most improbable. The use and purpose of the shell of the argonaut will be better understood if I briefly describe what I have witnessed of the treatment of its eggs by its near relative, the octopus.

"The eggs of the octopus," as I have elsewhere said, "when first laid, are small, oval, translucent granules, resembling little grains of rice, not quite an eighth of an inch long. They grow along and around a common stalk, to which every egg is separately attached, as grapes form part of a bunch. Each of the elongated bunches is affixed by a glutinous secretion to the surface of a rock or stone (never to seaweed, as has been erroneously stated), and hangs pendent by its stalk in a long white cluster, like a magnified catkin of the filbert, or, to use Aristotle's simile, like the fruit of the white alder. The length and number of these bunches varies according to the size and condition of

---

* Appendix to Sir Edward Belcher's 'Voyage of the "Samarang,"' by Mr. Arthur Adams, assistant surgeon to the expedition.

the parent. Those produced by a small octopus are seldom more than about three inches long, and from twelve to twenty in number; but a full-grown female will deposit from forty to fifty of such clusters, each about five inches in length. I have counted the eggs of which these clusters are composed, and find that there are about a thousand in each: so that a large octopus produces in one laying, usually extended over three days, a progeny of from 40,000 to 50,000. I have seen an octopus, when undisturbed, pass one of her arms beneath the hanging bunches of her eggs, and, dilating the membrane on each side of it into a boat-shaped hollow, gather and receive them in it as in a trough or cradle which exhibited in its general shape and outline a remarkable similarity to the shell of the argonaut, with the eggs of which octopod its own are almost identical in form and appearance. Then she would caress and gently rub them, occasionally turning towards them the mouth of her flexible exhalent and locomotor tube, like the nozzle of a fireman's hose-pipe, so as to direct upon them a jet of the excurrent water. I believe that the object of this syringing process is to free the eggs from parasitic animalcules, and possibly to prevent the growth of conferva, which, I have found, rapidly overspreads those removed from her attention." *

It has been suggested that the syringing may be for the purpose of keeping the water surrounding the eggs well aerated; but this is evidently erroneous, for the water ejected from the tube has been previously deprived of its oxygen, and consequently of its health-giving properties, whilst passing over the gills of the parent. Week after week, for fifty days, a brooding octopus will continue to attend to her eggs with the most watchful and assiduous

* 'The Octopus,' 1873, p. 57.

care, seldom leaving them for an instant except to take food, which, without a brief abandonment of her position, would be beyond her reach. Aristotle asserted that while the female is incubating she takes no food. This is incorrect; but in every case of the kind that has come under my observation the mother octopod, whenever she has been obliged to leave her nest, has returned to it as quickly as possible; and so I believe can, and does, the female argonaut to her shell, and that, too, without any difficulty. In her case the numerous clusters of eggs are all united at their origin to one slender and tapering stalk

FIG. 29.—THE PAPER NAUTILUS (*Argonauta argo*) CRAWLING.

which is fixed by a spot of glutinous matter to the body-whorl of the spiral shell.

This "paper-sailor," then, whom the poets have regarded as endowed with so much grace and beauty, and living in luxurious ease, is but a fine lady octopus after all. Turn her out of her handsome residence, and, instead of the fairy skimmer of the seas, you have before you an object apparently as free from loveliness and romance as her sprawling, uncanny-looking, relative. Instead of floating in her pleasure boat over the surface of the sea, the argonaut ordinarily crawls along the bottom, carrying her shell above her, keel uppermost; and the broad extremities of the two arms are not hoisted as sails, nor allowed when

THE "SAILING" OF THE NAUTILUS. 87

at rest to dangle over the side of the "boat;" but are used as a kind of hood by which the animal retains the shell in its proper position, as a man bearing a load on his shoulders holds it with his hands. When she comes to the surface, or progresses by swimming instead of walking, she does so in the same manner as the octopus: namely, by the forcible expulsion of water from her funnel-like tube.

But if truth compels us to deprive her of the counterfeit halo conferred on her by poets, we can award her, on behalf of science, a far nobler crown; namely, that of the Queen of the whole great Invertebrate Animal Kingdom. For, the *Cephalopoda*, of which the argonaut is a highly

FIG. 30.—THE PAPER NAUTILUS (*Argonauta argo*) SWIMMING.

organised member, are not only the highest in their own division, the *Mollusca*, but they are as far superior to all other animals which have no backbones, as man stands lord and king over all created beings that possess them.

Although in outward shape the spiral shell of the Pearly Nautilus (*Nautilus pompilius*) somewhat resembles that of the argonaut, its internal structure is very different. A section of it shows that it is divided into several chambers, each of which is partitioned off from the adjoining ones, the last formed or external one, in which the animal lives, being much larger than the rest. The object and mode of construction of these chambers is as follows. As the animal grows, a constant secretion of new material takes place on the edge of the shell. By this unceasing process of the addition of new shell in the form of a circular curve or coil around the older portion, the whole rapidly increases in size, both in diameter, and in the length of the chamber. The Nautilus, requiring to keep the secreting portion of its mantle applied to the lip of the shell, finds the chamber in which it dwells gradually becoming inconveniently long for it, and therefore builds up a wall behind itself, and continues its work of enlarging its premises in front. Each of these walls, concave in front, towards the mouth of the shell, and

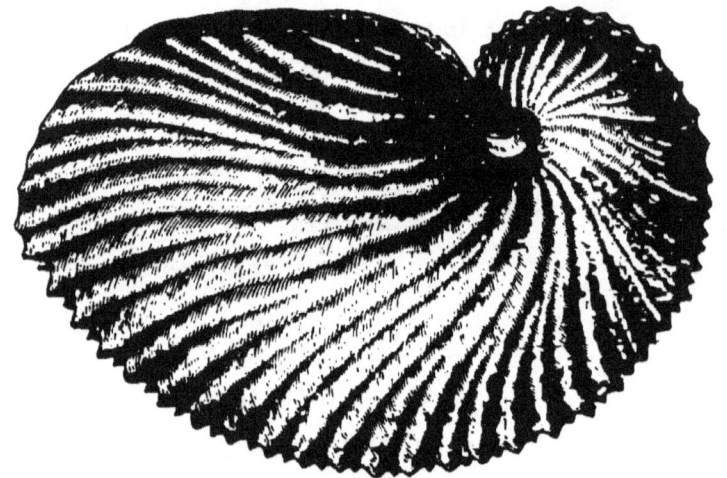

FIG. 31.—SHELL OF THE PAPER NAUTILUS (*Argonauta argo*).

concave behind, acts as a strong girder and support of the arch of the shell against the inward pressure of deep water: and it was formerly supposed that each successive chamber so constructed and vacated remained filled with air, and *thus* became an additional float by which the constantly increasing weight of the growing shell was counter-balanced. By this beautiful adjustment of augmented floating power to increased weight, the buoyancy of the shell would be secured and its specific gravity maintained as nearly as possible equal to that of the surrounding water. This adjustment does

THE "SAILING" OF THE NAUTILUS. 89

probably take place, but in a somewhat different manner. As the Nautilus inhabits a depth of from twenty to forty fathoms, it is evident that the air within its shell would be displaced by the pressure of such a column of water.* Accordingly, in every instance of the capture of a Nautilus the chambers of its shell have been found filled with water. It is not improbable that the fluid they contain may be less compressed, and exert less pressure from within outwards

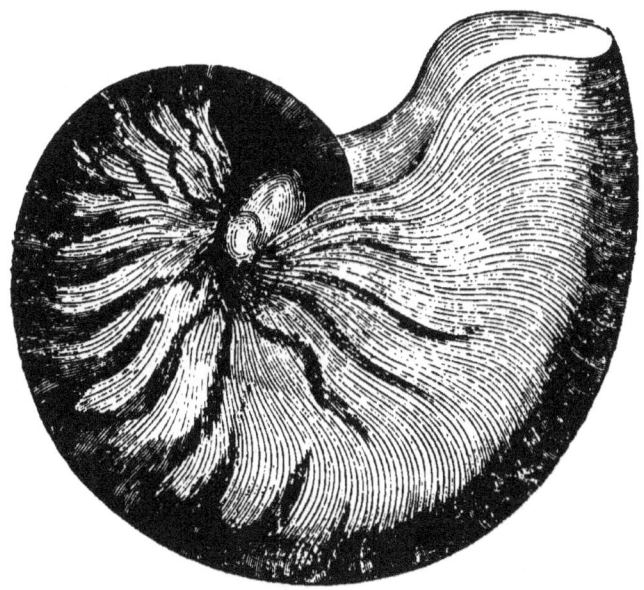

FIG. 32.—SHELL OF THE PEARLY NAUTILUS (*Nautilus pompilius*).

than that of the external superincumbent column of water, and that by this unbalanced pressure—under the same

* "At 100 fathoms the pressure exceeds 265 lbs. to the square inch. Empty bottles, securely corked, and sunk with weights beyond 100 fathoms, are always crushed. If filled with liquid the cork is driven in, and the liquid replaced by salt water; and in drawing the bottle up again the cork is returned to the neck of the bottle, generally in a reversed position."—Sir F. Beaufort, quoted by Dr. S. P. Woodward in his 'Manual of the Mollusca.'

hydro-dynamic law which governs its mode of self-propulsion when swimming, and possibly in some degree within the control of the animal—the latter is relieved of much of the weight of its shell. When the Nautilus is at the bottom of the sea its movement is like that of a snail crawling along

FIG. 33.—THE PEARLY NAUTILUS (*Nautilus pompilius*), AND SECTION OF ITS SHELL. *After Professor Owen.*

a a, Partitions; *b b*, chambers; *b'*, the last-formed chamber, in which the animal lives; *c c*, the siphuncle; *d*, attaching muscle; *e e*, the hollow arms; *ff*, retractile tentacles; *g*, muscular disk, or foot; *h*, the eye; *i*, position of funnel.

upon the ground with its shell above it. The shell, in proportion to the size of the animal that inhabits it, is a heavy one, and unless it were rendered semi-buoyant, its owner's strength would be severely taxed by the effort to drag it along. By the means indicated this portable domicile is borne lightly above the body of the Nautilus, without in any way impeding its progress.

The chambers are all connected by a membranous tube slightly coated with nacre, which is connected with a large sac in the body of the animal, near the heart, and passes through a circular orifice and a short projecting tube in the centre of each partition wall, till it ends in the smallest chamber at the inner extremity of the shell. Dean Buckland believed this "syphon" to be an hydraulic apparatus acting as a "fine adjustment" of the specific gravity of the shell, by admitting water within it when expanded, and excluding it when contracted. As it contains an artery and vein near its origin at the mantle, Professor Owen has regarded it as subservient to the maintenance of a low vitality in the vacated portion of the shell. Dr. Henry Woodward is of the opinion that, whilst in the early life of the Nautilus this siphuncle forms the main point of attachment between the animal and its shell, it is in the adult "simply an aborted embryonal organ whose function is now filled by the shell-muscles, but which in the more ancient and straight-shelled representatives of the group (the Orthoceratites) was not merely an embryonal but an important organ in the adult."

Every one knows the shell of the Pearly Nautilus. It may be purchased at any shell-shop in a seaside watering-place, and is imported by hundreds every year from Singapore.* It is abundant in the waters of the Indian Archipelago, especially about the Molucca and Philippine Islands, and on the shores of New Caledonia and the Fiji

---

* I need hardly say that before the nacreous layer of the shell from which this animal takes its name is made visible, an outer deposit of dense calcareous matter has to be removed by hydrochloric acid : the pearly surface thus exposed is then easily polished.

and Solomon Islands. It has also been found alive on Pemba Island, near Zanzibar. It seems strange, therefore, that until about half a century ago hardly anything was known of the animal that secretes and inhabits it. Rumphius, a Dutch naturalist, in his 'Rarities of Amboyna,' published, in 1705, a description of one with an engraving, incorrect in drawing, and deficient in detail; and until 1832 this was the only information which existed concerning it. The great Cuvier never saw one, and being acquainted only with the two-gilled cephalopods, he regarded the head-footed mollusks as absolutely isolated from all other animals in the kingdom of nature, even from the other classes of the mollusca. It seemed, however, to Professor Owen, then only nineteen years of age, that in the only living representative of the four-gilled order, *Nautilus pompilius*, might be found the "missing link." When, therefore, in the year 1824, his fellow-student, Mr. George Bennett, was about to sail from England to the Polynesian Islands, young Richard Owen earnestly charged his friend to do his utmost to obtain, and bring home in alcohol, a specimen of the much-coveted Pearly Nautilus. The opportunity did not occur till one warm and calm Monday evening, the 24th of August, 1829, when a living Nautilus was seen at the surface of the water not far distant from the ship, in Marekini Bay, on the south-west coast of the Island of Erromango, New Hebrides, in the South Pacific Ocean. It looked like a dead tortoise-shell cat, as the sailors said. As it began to sink as soon as it was observed, it was struck at with a boat-hook, and was thus so much injured that it died shortly after being taken on board the ship. The shell was destroyed, but the soft body of the animal was preserved in spirits, and great was the joy of Mr. Owen when, in July, 1831, Mr. Bennett

THE "SAILING" OF THE NAUTILUS. 93

arrived with it in England, and presented it to the Royal College of Surgeons. Mr. Owen was then Assistant-Conservator of the Museum of the College under Mr. Clift, who was afterwards his father-in-law. He immediately commenced to anatomise, describe, and figure his rare acquisition, and in the early part of 1832 published the result of his work in the form of a masterly treatise, which proved to be the foundation of his future fame.*

Mr. Owen's investigations confirmed his previous supposition that the Pearly Nautilus is inferior in its organisation to octopus, sepia, or any other known cephalopod; that it is not isolated, but that it recedes towards the gasteropods, to which belong the snail, the periwinkle, &c., and that in some of its characters its structure is analo-

* It is so interesting to most of us to know something of the early work of our greatest men, and of the tide in their affairs, which, taken at the flood, led on to fortune, that I hope I may be excused for referring to the period when the distinguished chief of the Natural History Department of the British Museum, the great comparative anatomist, the unrivalled palæontologist, the illustrious physiologist, the venerable and venerated friend of all earnest students, was beginning to attract the attention, and to receive the approbation of his seniors as a promising young worker. In Messrs. Griffith and Pidgeon's Supplement to Cuvier's 'Mollusca and Radiata,' published in 1834, the treatise in question is thus mentioned : " We have much pleasure in referring to a most excellent memoir on *Nautilus pompilius*, by Mr. Owen, with elaborate figures of the animal, its shell, and various parts, published by direction of the Council of the College of Surgeons. The reader will find the most satisfactory information on the subject, and the scientific public will earnestly hope that the present volume will be the first of a similar series." This hope has been more than fulfilled. Dean Buckland, in his ' Bridgewater Treatise,' wrote of this work : " I rejoice in the present opportunity of bearing testimony to the value of Professor Owen's highly philosophical and most admirable memoir—a work not less creditable to the author than honourable to the Royal College of Surgeons, under whose auspices the publication has been so handsomely conducted."

gously related to the still lower *annulosa*, or worms. Mr. Owen was just about to start for Paris with the intention of presenting a copy of his book to his celebrated contemporary and friend, and of showing him his dissections of the Nautilus which had been the subject of his research, when he heard of Baron Cuvier's death. It must have been to him a great sorrow and a grievous disappointment.

The Pearly Nautilus, then, is a true cephalopod, in that it has its foot divided and arranged in segments around its head, but the form and number of these segments are very different from those of any other of its class. Instead of there being eight, as in the argonaut and octopus, or ten, as in sepia and the calamaries, the Nautilus has about ninety projecting in every direction from around the mouth. They are short, round, and tapering, of about the length and thickness of the fingers of a child. Some of them are retractile into sheaths, and they are attached to fleshy processes (which might represent the child's hand), overlying each other, and covering the mouth on each side. They have none of the suckers with which the arms and tentacles of all the other cuttles are furnished, but their annulose structure, like the rings of an earthworm's body, gives them some little prehensile power. None of these numerous fingerlike segments of the foot are flattened out like the broad membranous expansions of the argonaut, and, in fact, the Nautilus is without any members which can possibly be regarded as sails to hoist, or as oars with which to row. It has a strong beak, like the rest of the cuttles; but it has no ink-sac, for its shell is strong enough to afford it the protection which its two-gilled relatives have to seek in concealment.

The Pearly Nautilus usually creeps, like a snail, along the bed of the sea. It lives at the bottom, and feeds at the bottom, principally on crabs; and, as Dr. S. P. Woodward says, in his 'Manual of the Mollusca,' "perhaps often lies in wait for them, like some gigantic sea-anemone, with outspread tentacles." The shape of its shell is not well adapted for swimming, but it can ascend to the surface, if it so please, in the same manner as can all the cuttles—namely, by the outflow of water from its locomotor tube. The statement that it visits the surface of the sea of its own accord is at present, however, unconfirmed by observation.

But, if the Pearly Nautilus is the inferior and poor relation of the argonaut, it lives in a handsome house, and comes of an ancient lineage. The Ammonites, whose beautiful whorled and chambered shells, and the casts of them, are so abundant in every stratum, especially in the lias, the chalk, and the oolite, had four gills also. These Ammonites and the Nautili were amongst the earliest occupants of the ancient deep; and, with the Hamites, Turrilites, and others, lived upon our earth during a great portion of the incalculable period which has elapsed since it became fitted for animal existence, and in their time witnessed the rise and fall of many an animal dynasty. But they are gone now; and only the fossil relics of more than two thousand species (of which 188 were Nautili) remain to tell how important a race they were amongst the inhabitants of the old world seas. They and their congeners of the chambered shells, however, left one representative which has lived on through all the changes that have taken place on the surface of this globe since they became extinct—namely, *Nautilus pompilius*, the Nautilus of the pearly shell—the last of the Tetrabranchs.

I need offer no apology for endeavouring to explain the difference between the Nautilus of the chambered shell and the argonaut with the membranous arms which it was supposed to use as sails, when Webster, in his great standard dictionary, describes the one and figures the other as one and the same animal; and when a writer of the celebrity of Dr. Oliver Wendell Holmes also blends the two in the following poem, containing a sentiment as exquisite as its science is erroneous. I hope the latter distinguished and accomplished author, whose delightful writings I enjoy and highly appreciate, will pardon my criticism. I admit that the beauty of the thought might well atone for its inaccuracy, (of which the author is conscious,) were it not that the latter is made so attractive that truth appears harsh in disturbing it.

"THE CHAMBERED NAUTILUS."

"This is the ship of pearl, which poets feign
      Sails the unshadowed main,
      The venturous bark that flings
On the sweet summer wind its purpled wings,
In gulfs enchanted, where the siren sings,
      And coral reefs lie bare,
Where the cold sea-maids rise to sun their streaming hair.

Its webs of living gauze no more unfurl,
      Wrecked is the ship of pearl!
      And every chambered cell,
Where its dim, dreaming life was wont to dwell,
As the frail tenant shaped his growing shell,
      Before thee lies revealed,
Its irised ceiling rent, its sunless crypt unsealed!

Year after year beheld the silent toil
      That spread his lustrous coil;
      Still, as the spiral grew,
He left the past year's dwelling for the new,
Stole with soft step its shining archway through,
      Built up its idle door,
Stretched in his last-found home, and knew the old no more.

Thanks for the heavenly message brought by thee,
        Child of the wandering sea,
        Cast from her lap forlorn!
From the dead lips a clearer note is born
Than ever Triton blew from wreathèd horn!
        While on mine ear it rings,
Through the deep caves of thought I hear a voice that sings :—

'Build thee more stately mansions, O my soul,
        As the swift seasons roll!
        Leave thy low vaulted past;
Let each new temple, nobler than the last.
Shut thee from heaven with a dome more vast,
        Till thou at length art free,
Leaving thine outgrown shell by life's unresting sea.'"

BARNACLE GEESE—GOOSE BARNACLES.

THE belief that some wild geese, instead of being hatched from eggs, like other birds, grew on trees and rotten wood has never been surpassed as a specimen of ignorant credulity and persistent error.

There are two principal versions of this absurd notion. One is that certain trees, resembling willows, and growing always close to the sea, produced at the ends of their branches fruit in form like apples, and each containing the embryo of a goose, which, when the fruit was ripe, fell into the water and flew away. The other is that the geese were bred from a fungus growing on rotten timber floating at sea, and were first developed in the form of worms in the substance of the wood.

When and whence this improbable theory had its origin is uncertain. Aristotle does not mention it, and consequently Pliny and Ælian were deprived of the pleasure they would have felt in handing down to posterity, without investigation or correction, a statement so surprising. It is, comparatively, a modern myth; although we find that it was firmly established in the middle of the twelfth century, for Gerald de Barri, known in literature as Giraldus Cambrensis, mentions it in his 'Topographia Hiberniæ,' published in 1187. Giraldus, who was Archdeacon of Brecknock in the reign of Henry II., and tried hard, more than once, for the bishopric of St. David's, the functions of which he had temporarily administered without obtaining the title, was a vigorous and zealous reformer of Church abuses. Amongst the laxities of discipline against which he found it necessary to protest was the custom then prevailing of eating these Barnacle geese during Lent, under the plea that their flesh was not that of birds, but of fishes. He writes :—

"There are here many birds which are called Bernacæ, which nature produces in a manner contrary to nature, and very wonderful. They are like marsh-geese but smaller. They are produced from fir-timber tossed about at sea, and are at first like geese upon it. Afterwards they hang down by their beaks, as if from a sea-weed attached to the wood, and are enclosed in shells that they may grow the more freely. Having thus, in course of time, been clothed with a strong covering of feathers, they either fall into the water, or seek their liberty in the air by flight. The embryo geese derive their growth and nutriment from the moisture of the wood or of the sea, in a secret and most marvellous manner. I have seen with my own eyes more than a thousand minute bodies of these birds hanging from one piece of timber on the shore, enclosed in shells and already formed. Their eggs are not impregnated *in coitu*, like those of other birds, nor does the bird sit upon its eggs to hatch them, and in no corner of the world have they been known to build a nest. Hence the bishops and clergy in some parts of Ireland are in the habit of partaking of these birds on fast days, without scruple. But in doing so they are led into sin. For, if any one were to eat of the leg of our first parent, although he (Adam) was not born of flesh, that person could not be adjudged innocent of eating flesh."

This fable of the geese appears, however, to have been current at least a hundred years before Giraldus wrote, for Professor Max Müller, who treats of it in one of his "Lectures on the Science of Language," amongst many interesting references there given, quotes a Cardinal of the eleventh century, Petrus Damianus, who clearly describes, that version of it which represents the birds as bursting, when fully fledged, from fruit resembling apples.

It is a curious fact that these Barnacle geese have

troubled the priesthood of more than one creed as to the instructions they should give to the laity concerning the use of them as food. The Jews—all those, at least, who maintain a strict observance of the Hebrew Law—eat no meat but that of animals which have been slaughtered in a certain prescribed manner; and a doubt arose amongst them at the period we refer to, whether these geese should be killed as flesh or as fish. Professor Max Müller cites Mordechai,* as asking whether these birds are fruits, fish, or flesh; that is, whether they must be killed in the Jewish way, as if they were flesh. Mordechai describes them as birds which grow on trees, and says, "the Rabbi Jehuda, of Worms (who died 1216) used to say that he had heard from his father, Rabbi Samuel, of Speyer (about 1150), that Rabbi Jacob Tham, of Ramerü (who died 1171), the grandson of the great Rabbi Rashi (about 1140), had decided that they must be killed as flesh."

Pope Innocent III. took the same view; for at the Lateran Council, in 1215, he prohibited the eating of Barnacle geese during Lent. In 1277, Rabbi Izaak, of Corbeil, determined to be on the safe side, forbade altogether the eating of these birds by the Jews, "because they were neither flesh nor fish."

Michael Bernhard Valentine,† quoting Wormius, says that this question caused much perplexity and disputation amongst the doctors of the Sorbonne; but that they passed an ordinance that these geese should be classed as fishes, and not as birds; and he adds, that in consequence of this decision large numbers of these birds were annually sent to Paris from England and Scotland, for consumption in

---

* Riva, 1559, leaf 142*.
† 'Historia Simplicium,' lib. iii. p. 327.

Lent. Sir Robert Sibbald* refers to this, and says that Normandy was the locality from which the French capital was reported to be principally supplied; but that in fact the greater number of these geese came from Holland. The date of this edict is not given.

Professor Max Müller says that in Brittany, Barnacle geese are still allowed to be eaten on Fridays, and that the Roman Catholic Bishop of Ferns may give permission to people out of his diocese to eat these birds at his table.

In Bombay, also, where fish is prohibited as food to some classes of the population, the priests call this goose a "sea-vegetable," under which name it is allowed to be eaten.

Various localities were mentioned as the breeding-places of these arboreal geese. Gervasius of Tilbury,† writing about 1211, describes the process of their generation in full detail, and says that great numbers of them grew in his time upon the young willow trees which abounded in the neighbourhood of the Abbey of Faversham, in the county of Kent, and within the Archiepiscopate of Canterbury. The bird was there commonly called the *Barneta*.

Hector Boethius, or Boece, the old Scottish historian, combats this version of the story. His work, written in Latin, in 1527, was translated into quaint Scottish in 1540, by John Bellenden, Archdeacon of Murray. In his fourteenth chapter, "Of the nature of claik geis, and of the syndry maner of thair procreatioun, And of the ile of Thule," he says:—

"Restis now to speik of the geis generit of the see namit clakis. Sum men belevis that thir clakis growis on treis be the nebbis. Bot thair opinioun is vane. And becaus the nature and procreatioun of thir clakis is strange we have maid na lytyll laubore and deligence to

---

* Prodrom. Hist. Nat. Scot. parts 2, lib. iii. p. 21, 1684.
† Otia Imperialia, iii. 123.

serche ye treuth and verite yairof, we have salit throw ye seis quhare thir clakis ar bred, and I fynd be gret experience, that the nature of the seis is mair relevant caus of thir procreatioun than ony uther thyng."

From the circumstances attending the finding of "ane gret tree that was brocht be alluvion and flux of the see to land, in secht of money pepyll besyde the castell of Petslego, in the yeir of God ane thousand iiii. hundred lxxxx, and of a see tangle hyngand full of mussill schellis," brought to him by "Maister Alexander Galloway, person of Kynkell," who knowing him to be "richt desirus of sic uncouth thingis came haistely with the said tangle," he arrives at the conclusion, by a process of reasoning highly satisfactory and convincing to himself, that,

"Be thir and mony othir resorcis and examplis we can not beleif that thir clakis ar producit be ony nature of treis or rutis thairof, but allanerly be the nature of the Oceane see, quhilk is the caus and production of mony wonderful thingis. And becaus the rude and ignorant pepyl saw oftymes the fruitis that fel of the treis (quhilkis stude neir the see) convertit within schort tyme in geis, thai belevit that thir geis grew apon the treis hingand be thair nebbis sic lik as appillis and uthir frutis hingis be thair stalkis, bot thair opinioun is nocht to be sustenit. For als sone as thir appillis or frutis fallis of the tre in the see flude thay grow first wormeetin. And be schort process of tyme are alterat in geis."

In describing the bird thus produced, Boethius declares that the male has a sharp, pointed beak, like the gallinaceous birds, but that in the female the beak is obtuse as in other geese and ducks.

According to other authors, this wonderful production of birds from living or dead timber was not confined to England and Scotland. Vincentius Bellovacensis [*] (1190–

---

[*] For this quotation and the following one I am indebted to Professor Max Müller's Lecture before referred to.

1264) in his 'Speculum Naturæ,' xvii. 40, states that it took place in Germany, and Jacob de Vitriaco (who died 1244) mentions its occurrence in certain parts of Flanders.

Jonas Ramus gives a somewhat different version of the process as it occurs in Norway. He writes:* "It is said that a particular sort of geese is found in Nordland, which leave their seed on old trees, and stumps and blocks lying in the sea; and that from that seed there grows a shell fast to the trees, from which shell, as from an egg, by the heat of the sun, young geese are hatched, and afterwards grow up; which gave rise to the fable that geese grow upon trees."

But, strange to say, if any painstaking enquirer, wishing to investigate the matter for himself, went to a locality where it was said the phenomenon regularly occurred, he was sure to find that he had literally, "started on a wild-goose chase," and had come to the wrong place. This was the experience of Æneas Sylvius Piccolomini, afterwards Pope Pius II., who complained that miracles will always flee farther and farther away; for when he was on a visit (about 1430) to King James I., of Scotland,† and enquired after the tree which he most eagerly desired to see, he was told that it grew much farther north, in the Orkney Islands.

Notwithstanding the suspicious fact that the prodigy receded like Will o' the Wisp, whenever it was persistently followed up, Sebastian Munster, who relates ‡ the foregoing

* 'Chorographical Description of Norway,' p. 244.
† Æneas Sylvius gives us information concerning the personal appearance of his royal host, whom he describes as, "*hominem quadratum et multa pinguedine gravem*,"—literally, " a square-built man, heavy with much fat."
‡ 'Cosmographia Universalis,' p. 49, 1572.

anecdote of Æneas Sylvius, appears to have entertained no doubt of the truth of the report, for he writes:—

FIG. 34.—THE GOOSE TREE. *Copied from Gerard's 'Herball,' 1st edition.**

" In Scotland there are trees which produce fruit, conglomerated of

---

* The original of this picture is a small wood-cut in Matthias de Lobel's 'Stirpium Historia,' published in 1870. The birds within the shells were added by Gerard. Aldrovandus, in copying it, gave leaves to the tree, as shown on page 110.

their leaves; and this fruit, when in due time it falls into the water beneath it, is endowed with new life, and is converted into a living bird, which they call the 'tree-goose.' This tree grows in the Island of Pomonia, which is not far from Scotland, towards the north. Several old cosmographers, especially Saxo Grammaticus, mention the tree, and it must not be regarded as fictitious, as some new writers suppose."

Julius Cæsar Scaliger* (1540) gives another reading of the legend, in which it is asserted that the leaves which fall from the tree into the water are converted into fishes, and those which fall upon the land become birds.

Thus this extraordinary belief held sway, and remained strong and invincible, although from time to time some man of sense and independent thought attempted to turn the tide of popular error. Albertus Magnus (who died 1280) showed its absurdity, and declared that he had seen the bird referred to lay its eggs and hatch them in the ordinary way. Roger Bacon (who died in 1294) also contradicted it, and Belon, in 1551, treated it with ridicule and contempt. Olaus Wormius † seems to have believed in it, though he wrote cautiously about it. Olaus Magnus (1553) mentions it, and apparently accepts it as a fact, occurring in the Orkneys, on the authority of "a Scotch historian who diligently sets down the secrets of things," and then dismisses it in three lines.

Passing over many other writers on the subject, we come to the time of the reign of Queen Elizabeth, when (in 1597) "John Gerarde, Master in Chirurgerie, London," published his "Herball, or Generall Historie of Plants gathered by him," and in the last chapter thereof solemnly declared, that he had actually witnessed the transformation of "certaine shell fish" into Barnacle Geese, as follows.

* Exercit. 59, sect. 2.   † 'Museum,' p. 257.

Of the Goose tree, Barnacle tree, or the tree bearing Geese.

Britanicæ Conchæ anatifera.

THE BREED OF BARNACLES.

¶ *The Description.*

Hauing trauelled from the Grasses growing in the bottome of the fenny waters, the Woods, and mountaines, euen vnto Libanus itselfe; and also the sea, and bowels of the same, wee are arriued at the end of our History; thinking it not impertinent to the conclusion of the same, to end with one of the maruels of this land (we may say of the World). The history whereof to set forth according to the worthinesse and raritie thereof, would not only require a large and peculiar volume, but also a deeper search into the bowels of Nature, then my intended purpose will suffer me to wade into, my sufficiencie also considered; leauing the History thereof rough hewen, vnto some excellent man, learned in the secrets of nature, to be both fined and refined; in the meane space take it as it falleth out, the naked and bare truth, though vnpolished. There are found in the North parts of Scotland and the Islands adiacent, called Orchades, certaine trees whereon do grow certaine shells of a white colour tending to russet, wherein are contained little liuing creatures: which shells in time of maturity doe open, and out of them grow those little liuing things, which falling into the water do become fowles, which we call Barnacles; in the North of England, brant Geese; and in Lancashire, tree Geese: but the other that do fall vpon the land perish and come to nothing. Thus much by the writings of others, and also from the mouthes of people of those parts, which may very well accord with truth.

But what our eies haue seene, and hands haue touched we shall declare. There is a small Island in Lancashire, called the Pile of Foulders, wherein are found the broken pieces of old and bruised ships some whereof haue beene cast thither by shipwracke, and also the trunks and bodies with the branches of old and rotten trees, cast vp there likewise; whereon is found a certaine spume or froth that in time breedeth vnto certaine shells, in shape like those of the Muskle, but sharper pointed, and of a whitish colour ; wherein is contained a thing in forme like a lace of silke finely wouen as it were together, of a whitish colour, one end whereof is fastened vnto the inside of the shell, euen as the fish of Oisters and Muskles are : the other end is made fast vnto the belly of a rude masse or lumpe, which in time commeth to the shape and forme of a Bird : when it is perfectly formed the shell gapeth open, and the first thing that appeareth is the foresaid lace or string ; next come the legs of the bird hanging out, and as it groweth greater it openeth the shell by degrees, til at length it is all come forth, and hangeth onely by the bill : in short space after it commeth to full maturitie, and falleth into the sea, where it gathereth feathers, and groweth to a fowle bigger than a Mallard, and lesser than a Goose, hauing blacke legs and bill or beake, and feathers blacke and white, spotted in such manner as is our Magpie, called in some places a Pie-Annet, which the people of Lancashire call by no other name than a tree Goose : which place aforesaid, and all those parts adjoyning do so much abound therewith, that one of the best is bought for three pence. For the truth hereof, if any doubt, may it please them to repaire vnto me, and I shall satisfie them by the testimonie of good witnesses.

Moreover, it should seeme that there is another sort hereof; the History of which is true, and of mine owne knowledge; for trauelling vpon the shore of our English coast betweene Douer and Rumney, I found the trunke of an old rotten tree, which (with some helpe that I procured by Fishermen's wiues that were there attending their husbands' returne from the sea) we drew out of the water vpon dry land ; vpon this rotten tree I found growing many thousands of long crimson bladders, in shape like vnto puddings newly filled, before they be sodden, which were very cleere and shining ; at the nether end whereof did grow a shell fish, fashioned somewhat like a small Muskle, but much whiter, resembling a shell fish that groweth vpon the rockes about Garnsey and Garsey, called a Lympit : many of these shells I brought with me to London, which after I had opened I found in them liuing things without forme or shape ; in others which were neerer come to ripenesse I found liuing things that were very naked, in shape like a Bird : in others, the Birds couered with soft downe, the shell halfe open, and the Bird ready to fall out, which no doubt were the Fowles called Barnacles. I dare not absolutely auouch euery circumstance of the first part of this history, concerning the tree that beareth those buds aforesaid, but will leaue it to a further consideration ; howbeit, that which I haue seene with mine eies, and handled with mine hands, I dare confidently auouch, and boldly put downe for verity. Now if any will object that this tree which I saw might be one of those before mentioned, which either by the waues of the sea or some violent wind had beene ouerturned as many other trees are; or that any trees falling into those seas about the Orchades, will of themselues bear the like Fowles, by reason of those seas and waters, these being so probable conjectures, and likely to be true, I may not without prejudice gainsay, or endeauour to confute.

¶ *The Place.*

The bordes and rotten plankes whereon are found these shels breeding the Barnakle, are taken vp in a small Island adioyning to Lancashire, halfe a mile from the main land, called the Pile of Foulders.

¶ *The Time.*

They spawn as it were in March and Aprill; the Geese are formed in May and June, and come to fulnesse of feathers in the moneth after.

And thus hauing through God's assistance discoursed somewhat at large of Grasses, Herbes, Shrubs, Trees, and Mosses, and certaine Excrescenses of the Earth, with other things moe, incident to the historie thereof, we conclude and end our present Volume, with this wonder of England. For the which God's name be euer honored and praised.

Gerard was probably a good botanist and herbalist; but Thomas Johnson, the editor of a subsequent issue of his book, tells us that

"He, out of a propense good will to the publique advancement of this knowledge, endeavoured to performe therein more than he could well accomplish, which was partly through want of sufficient learning; but," he adds, "let none blame him for these defects, seeing he was neither wanting in pains nor good will to performe what hee intended: and there are none so simple but know that heavie burthens are with most paines vndergone by the weakest men; and although there are many faults in the worke, yet iudge well of the Author; for, as a late writer well saith:—'To err and to be deceived is human, and he must seek solitude who wishes to live only with the perfect.'"

It is difficult to comply with the request to think well of one who, writing as an authority, deliberately promulgated, with an affectation of piety, that which he must have known to be untrue, and who was, moreover, a shameless plagiarist; for Gerard's ponderous book is little more than a translation of Dodonæus, whole chapters having been taken verbatim from that comparatively unread author without acknowledgment.

After this series of erroneous observations, self-delusion, and ignorant credulity, it is refreshing to turn to the pages of the two little thick quarto volumes of Gaspar Schott.* This learned Jesuit made himself acquainted with everything that had been written on the subject, and besides the authors I have referred to, quotes and compares the statements of Majolus, Abrahamus Ortelius, Hieronymus Cardanus, Eusebius, Nierembergius, Deusingius, Odoricus, Gerhardus de Vera, Ferdinand of Cordova, and many others. He then gives, firmly and clearly, his own opinion that the assertion that birds in Britain spring from the fruit or leaves of trees, or from wood, or from fungus, or from shells, is without foundation, and that neither reason, experience, nor authority tend to confirm it. He concedes that worms may be bred in rotting timber, and even that they may be of a kind that fly away on arriving at maturity (referring probably to caterpillars being developed into moths), but that birds should be thus generated, he says, is simply the repetition of a vulgar error, for not one of the authors whom he has examined has seen what they all affirm; nor are they able to bring forward a single eye-witness of it. He asks how it can be possible that animals so large and so highly-organised as these birds

* 'Physica Curiosa, sive Mirabilia Naturæ et Artis,' 1662, lib. ix. cap. xxii. p. 960.

can grow from puny animalcules generated in putrid wood. He further declares that these British geese are hatched from eggs like other geese, which he considers proved by the testimony of Albertus Magnus, Gerhardus de Vera, and of Dutch seamen, who, in 1569, gave their written declaration that they had personally seen these birds sitting on their eggs, and hatching them, on the coasts of Nova Zembla.

FIG. 35.—THE BARNACLE GOOSE TREE. *After Aldrovandus.*

In marked and disgraceful contrast with this careful and philosophical investigation and its author's just deductions from it, is 'A Relation concerning Barnacles by Sir Robert Moray, lately one of His Majesty's Council for the Kingdom of Scotland,' read before the Royal Society, and published in the 'Philosophical Transactions,' No. 137, January and February, 1677-8.

Describing "a cut of a large Firr-tree of about two and a half feet diameter, and nine or ten feet long," which he saw on the shore in the Western Islands of Scotland, and which had become so dry that many of the Barnacle shells with which it had been covered had been rubbed off, he says:—

"Only on the parts that lay next the ground there still hung multitudes of little Shells, having within them little Birds, perfectly

FIG. 36.—DEVELOPMENT OF BARNACLES INTO GEESE. *After Aldrovandus.*

shap'd, supposed to be Barnacles. The Shells hung very thick and close one by another, and were of different sizes. Of the colour and consistence of Muscle-Shells, and the sides and joynts of them joyned with such a kind of film as Muscle-Shells are, which serves them for a Hing to move upon, when they open and shut. . . . The Shells hang at the Tree by a Neck longer than the Shell, of a kind of Filmy substance, round, and hollow, and creased, not unlike the Wind-pipe of a chicken, spreading out broadest where it is fastened to the Tree, from which it seems to draw and convey the matter which serves for the growth and vegetation of the Shell and the little Bird within it. This Bird in every Shell that I opened, as well the least as the biggest, I found so curiously and compleatly formed, that there appeared nothing wanting as to internal parts, for making up a perfect Sea-fowl: every little part appearing so distinctly that the whole looked like a large Bird seen through a concave or diminishing glass, colour and feature being everywhere so clear and neat. The little Bill, like that of a Goose; the eyes marked; the Head, Neck, Breast, Wings, Tail, and Feet formed, the Feathers everywhere perfectly shap'd, and blackish coloured; and the Feet like those of other Water-fowl, to my best remembrance. All being dead and dry, I did not look after the internal parts of them. Nor did I ever see any of the little Birds alive, nor met with anybody that did. Only some credible persons have assured me they have seen some as big as their fist."

It seems almost incredible that little more than two hundred years ago this twaddle should not only have been laid before the highest representatives of science in the land, but that it should have been printed in their "Transactions" for the further delusion of posterity.

Ray, in his edition of Willughby's Ornithology, published in the same year as the above, contradicted the fallacy as strongly as Gaspar Schott; and (except that he incidentally admits the possibility of spontaneous generation in some of the lower animals, as insects and frogs) in language so similar that I think he must have had Schott's work before him when he wrote.

Aldrovandus[*] tells us that an Irish priest, named Octavianus, assured him with an oath on the Gospels that he had seen and handled the geese in their embryo condition; and he adds that he "would rather err with the majority than seem to pass censure on so many eminent writers who have believed the story."

In 1629 Count Maier (Michaelus Meyerus—these old authors when writing in Latin, latinized their names also)

[*] 'Ornithologia,' lib. xix. p. 173, ed. 1603.

published a monograph 'On the Tree-bird'* in which he explains the process of its birth, and states that he opened a hundred of the goose-bearing shells and found the rudiments of the bird fully formed.

> So slow Bootes underneath him sees,
> In th' icy isles, those goslings hatched on trees,
> Whose fruitful leaves, falling into the water,
> Are turned, they say, to living fowls soon after;
> So rotten sides of broken ships do change,
> To barnacles, O, transformation strange!
> 'Twas first a green tree; then a gallant hull;
> Lately a mushroom; then a flying gull.†

Now, let us turn from fiction to facts.

Almost every one is acquainted with at least one kind of the Barnacle shells which were supposed to enclose the

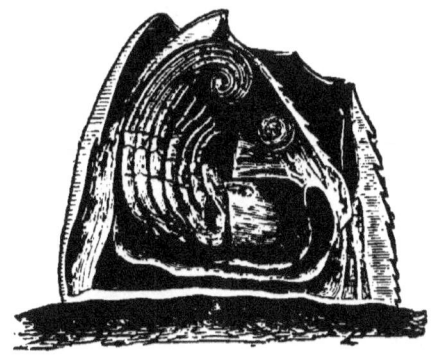

FIG. 37.—SECTION OF A SESSILE BARNACLE. *Balanus tintinnabulum.*

embryo of a goose, namely the small white conical hillocks which are found, in tens of thousands, adhering to stones, rocks, and old timber such as the piles of piers, and may be seen affixed to the shells of oysters and mussels in any fishmonger's shop. The little animals which secrete and

* 'De Volucri Arborea,' 1629.
† Du Bartas' "Divine Week" p. 228. Joshua Sylvester's translation.

inhabit these shells belong to a sub-class and order of the Crustacea, called the *Cirrhopoda*, because their feet (*poda*), which in the crab and lobster terminate in claws, are modified into tufts of curled hairs (*cirri*), or feathers. When the animal is alive and active under water, a crater may be seen to open on the summit of the little shelly mountain, and, as if from the mouth of a miniature volcano, there issue from this aperture, from between two inner shells, the *cirri* in the form of a feathery hand, which clutches at the water within its reach, and is then quickly retracted within the shell. During this movement the hair-fringed fingers have filtered from the water and conveyed towards the mouth within the shell, for their owner's nutriment, some minute solid particles or animalcules, and this action of the casting-net alternately shot forth and retracted continues for hours incessantly, as the water flows over its resting-place. The animal can live for a long time out of water, and in some situations thus passes half its life. Under such circumstances, the shells, containing a reserve of moisture, remain firmly closed until the return of the tide brings a fresh supply of water and food. These are the "acorn-barnacles," the *balani*, commonly known in some localities as "chitters."

Barnacles of another kind are those furnished with a long stem, or peduncle, which Sir Robert Moray described as "round, hollow, and creased, and not unlike the wind-pipe of a chicken." The stem has, in fact, the ringed formation of the annelids, or worms. The shelly valves are thin, flat, and in shape somewhat like a mitre. They are composed of five pieces, two on each side, and one, a kind of rounded keel along the back of the valves, by which these are united. The shells are delicately tinted with lavender or pale blue varied with white, and the edges are frequently of a bright chrome yellow or orange colour.

It is not an uncommon occurrence for a large plank entirely covered with these "necked barnacles" to be found floating at sea and brought ashore for exhibition at some watering-place; and I have more than once sent portions of such planks to the Aquaria at Brighton, and the Crystal Palace.

It is most interesting to watch a dense mass of living cirripedes so closely packed together that not a speck of

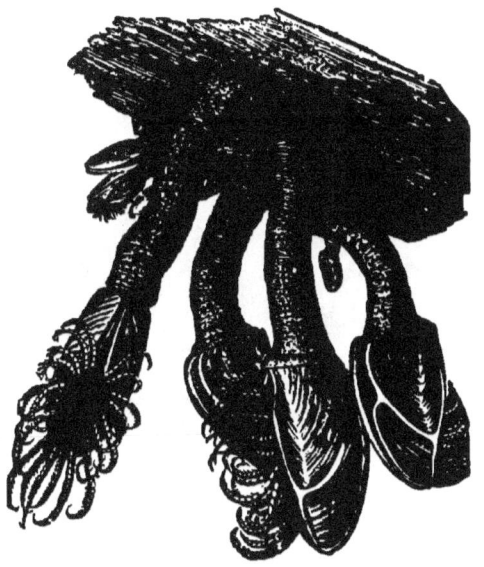

FIG. 38.—PEDUNCULATED BARNACLES. (*Lepas anatifera.*)

the surface of the wood is left uncovered by them; their fleshy stalks overhanging each other, and often attached in clusters to those of some larger individuals; their plumose casting-nets ever gathering in the food that comes within their reach, and carrying towards the mouth any solid particles suitable for their sustenance. How much of insoluble matter barnacles will eliminate from the water is shown by the rapidity with which they will render turbid sea water clear and transparent. The most common species of these "necked barnacles" bears the name of "*Lepas anatifera*," "the duck-bearing *Lepas*." It was so entitled by Linnæus, in recognition of its having been connected with the fable, which, of course, met with no credit from him.

Fig. 39 represents the figure-head of a ship, partly covered with barnacles, which was picked up about thirty miles off Lowestoft on the 22nd of October, 1857. It was described in the *Illustrated London News*, and the pro-

FIG. 39.—A SHIP'S FIGURE-HEAD WITH BARNACLES ATTACHED TO IT.

prietors of that paper have kindly given me a copy of the block from which its portrait was printed.

Others of the barnacles affix themselves to the bottoms of ships, or parasitically upon whales and sharks, and those of the latter kind often burrow deeply into the skin of their host. Fig. 40 is a portrait of a *Coronula diadema* taken from the nose of a whale stranded at Kintradwell, in the north of Scotland, in 1866, and sent to the late Mr. Frank Buckland. Growing on this *Coronula* are three of the curious eared barnacles, *Conchoderma aurita*, the *Lepas aurita* of Linnæus. The species of the whale from which these Barnacles were taken was not mentioned, but it was probably the "hunch-backed" whale, *Megaptera longimana*,

FIG. 40.—WHALE BARNACLE (*Coronula diadema*), WITH THREE *Conchoderma aurita* ATTACHED TO IT.

which is generally infested with this *Coronula*. This very illustrative specimen was, and I hope still is, in Mr. Buckland's Museum at South Kensington. It was described by him in *Land and Water*, of May 19th, 1866, and I am indebted to the proprietors of that paper for the accompanying portrait of it.

The young Barnacle when just extruded from the shell of its parent is a very different being from that which it will be in its mature condition. It begins its life in a form exactly like that of an entomostracous crustacean, and, like a Cyclops, has one large eye in the middle of its forehead. In this state it swims freely, and with great activity. It undergoes three moults, each time altering its figure, until at the third exuviation it has become enclosed in a

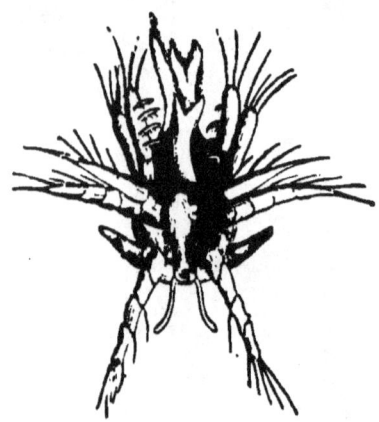

FIG. 41.—A YOUNG BARNACLE. (*Larva of Chthamalus stellatus.*)

bivalve shell, and has acquired a second eye. It is now ready to attach itself to its abiding-place; so, selecting its future residence, it presses itself against the wood, or whatever the substance may be, pours out from its two antennæ a glutinous cement, which hardens in water, and thus fastens itself by the front of its head, is henceforth a fixture for life, and assumes the adult form in which most persons know it best.*

* If any of my readers wish to observe the development of young barnacles they may easily do so. The method I have generally adopted has been as follows: Procure a shallow glass or earthenware milk-pan that will hold at least a gallon. Fill this to within an inch

It is unnecessary for me to describe more minutely the anatomy of the Cirripedes; I have said enough to show

of the top with sea-water, and place it in any shaded part of a room—not in front of a window. Put in the pan six or eight pebbles or clean shells of equal height, say 1½ or 2 inches, and on them lay a clean sheet of glass, which, by resting on the pebbles, is brought to within about 2½ inches of the surface of the water. Select some limpets or mussels having acorn-barnacles on them; carefully cut out the limpet or mussel, and clean nicely the interior of the shell; then place a dozen or more of these shells on the sheet of glass, and the barnacles upon them will be within convenient reach of any observation with a magnifying glass. If this be done in the month of March, the experimenter will not have to wait long before he sees young *Balani* ejected from the summits of some of the shells. Up to the moment of their birth each of them is inclosed in a little cocoon or case, in shape like a canary-seed, and most of them are tossed into the world whilst still enclosed in this. In a few seconds this casing is ruptured longitudinally, apparently by the struggles of its inmate, which escapes at one end, like a butterfly emerging from its chrysalis, and swims freely to the surface of the water, leaving the split cocoon or case at the bottom of the pan. Some few of the young barnacles seem to be freed from the cocoon before, or at the moment of, extrusion. From three to a dozen or more of these escape with each protrusion of the cirri of the parent, and as the parturient barnacle will put forth its feathery casting net at least twenty times in a minute for an hour or more, it follows that as many as ten thousand young ones may be produced in an hour. These, as they are cast forth at each pulsation of the parent's cirri, fall upon the clean sheet of glass, and may be taken up in a pipette, and placed under a microscope, or removed to a smaller vessel of sea-water, for minute and separate investigation. It seems strange that animals which, like the oyster and the barnacles, are condemned in their mature condition to lead so sedentary a life, should in the earlier stages of their existence swim freely and merrily through the water—young fellows seeking a home, and when they have found it, although their connubial life must be a very tame one, settling down, and not caring to rove about any more for the remainder of their days. These young *Balani* dart about like so many water-fleas, and yet, after a few days of freedom, they become fixed and immovable, the inhabitants of the pyramidal shells which grow in such abundance on other shells, stones, and old wood.

the nature of the plumose appurtenances which, hanging from the dead shells, were supposed to be the feathers of a little bird within; but it is difficult to understand how any one could have seen in the natural occupant of the shell, "the little bill, like that of a goose, the eyes, head, neck, breast, wings, tail, and feet, like those of other water-fowl," so precisely and categorically detailed by Sir Robert Moray. As Pontoppidan, who denounced the whole story, as being "without the least foundation," very truly says, "One must take the force of imagination to help to make it look so!"

As to the origin of the myth, I venture to differ entirely from philologists who attribute it to "language," and "a similarity of names," for, although, as Professor Max Müller observes in one of his lectures, "words without definite meanings are at the bottom of nearly all our philosophical and religious controversies," it certainly is not applicable in this instance. Every quotation here given shows that the mistake arose from the supposed resemblance of the plumes of the cirrhopod, and the feathers of a bird, and the fallacious deductions derived therefrom. The statements of Maier (p 112), Gerard (p. 106), Sir Robert Moray (p. 110), &c., prove that this fanciful misconception sprang from erroneous observation. The love of the marvellous inherent in mankind, and especially prevalent in times of ignorance and superstition, favoured its reception and adoption, and I believe that it would have been as widely circulated, and have met with equal credence, if the names of the cirripede and of the goose that was supposed to be its offspring had been far more dissimilar than, at first, they really were.

Setting aside several ingenious and far-fetched derivations that have been proposed, I think we may safely regard the word "barnacle," as applied to the cirrhopod, as a corruption of *pernacula*, the diminutive of *perna*, a bivalve mollusk, so-called from the similarity in shape of its shell to that of a ham—*pernacula* being changed to *bernacula*. In some old Glossaries *perna* is actually spelt *berna*.

To arrive at the origin of the word "barnacle," or "bernicle," as applied to the goose, we must understand that this bird, *Anser leucopsis*, was formerly called the "brent," "brant," or "bran" goose, and was supposed to be identical with the species, *Anser torquatus*, which is now known by that name. The Scottish word for "goose" is "clake," or "clakis,"* and I think that the suggestion made long ago to Gesner † (1558), by his correspondent, Joannes Caius, is correct, that the word "barnacle" comes from "branclakis," or "barnclake," "the dark-coloured goose."

Professor Max Müller is of the opinion that its Latin name may have been derived from *Hibernicæ, Hiberniculæ, Berniculæ*, as it was against the Irish bishops that Geraldus wrote, but I must say that this does not commend itself to me; for the name *Bernicula* was not used in the early times to denote these birds. Giraldus himself described them as *Bernacæ*, but they were variously known, also, as *Barliates, Bernestas, Barnetas, Barbates*, etc.

I agree with Dr. John Hill,‡ that " the whole matter that gave origin to the story is that the 'shell-fish' (cirripedes), supposed to have this wonderful production usually adhere to old wood, and that they have a kind of fibres hanging out of them, which, in some degree, resemble feathers of

---

* See the quotation from Hector Boetius, p. 101.
† 'Historia Animalium,' lib. iii. p. 110.
‡ 'History of Animals,' p. 422. 1752.

some bird. From this slight origin arose the story that they contained real birds: what grew on trees people soon asserted to be the fruit of trees, and, from step to step, the story gained credit with the hearers," till, at length, Gerard had the audacity to say that he had witnessed the transformation.

The Barnacle Goose is only a winter visitor of Great Britain. It breeds in the far north, in Greenland, Iceland, Spitzbergen, and Nova Zembla, and probably, also, along the shores of the White Sea. There are generally some specimens of this prettily-marked goose in the gardens of the Zoological Society in the Regent's Park, London; and they thrive there, and become very tame. In the months of December and January these geese may often be seen hanging for sale in poulterers' shops; and he who has tasted one well cooked may be pardoned if the suspicion cross his mind that the "monks of old," and "the bare-footed friars," as well as the laity, may not have been unwilling to sustain the fiction in order that they might conserve the privilege of having on their tables during the long fast of Lent so agreeable and succulent a "vegetable" or "fish" as a Barnacle Goose.

THE END.

www.ingramcontent.com/pod-product-compliance
Lightning Source LLC
Chambersburg PA
CBHW020810230426
43666CB00007B/944